Conceptual
Physical
Science
Explorations

Paul G. Hewitt
John Suchocki
Leslie A. Hewitt
Dean Baird

Laboratory Manual

Addison Wesley

San Francisco Boston New York
Capetown Hong Kong London Madrid Mexico City
Montreal Munich Paris Singapore Sydney Tokyo Toronto

Cover Credit: "Primal Motion"
        Michael T. Stewart ©/Molten Images.com

ISBN: 0-321-05183-1

    3 4 5 6 7 8 9 10 - MAL - 04 03 02

www.aw.com/physics

Introduction

Many students enter middle school and junior high with very few hands-on science learning experiences. The activities and experiments in this manual are designed to provide the experiences students need. Whether it's an attempt to use a battery and wires to get a bulb to light, collecting carbon dioxide by water displacement, or typing minerals based on observable characteristics, these explorations are designed to put students in direct interaction with the natural world in ways that will deepen both their curiosity and their understanding. It has been suggested that we retain about 5% of what we read and about 75% of what we do. So the activities and experiments in this manual play an important role in the complete *Conceptual Physical Science—Explorations* program of instruction. Enjoy!

Acknowledgments

Many of the physics experiments and activities in this manual originated with Paul (Pablo) Robinson, author of the lab manuals for both the high school and college versions of *Conceptual Physics*. So we owe a big thank you to Pablo.

Thanks to Earl R. Feltyberger, of Nicolet High School in Glendale, Wisconsin for the activity *Bouncy Board*.

Thanks to Ted Brattstrom for *Mystery Powders*. For *Salt and Sand*, we are grateful to Erwin W. Richter.

For contributions to nearly all the earth science labs, we are grateful to Leslie's husband, Bob Abrams (also a geologist). For valuable feedback and suggestions, we thank City College of San Francisco physics and geology instructor Jim Court. For other geology resources, we are grateful to the American Geological Institute and the National Association of Geology Teachers.

For astronomy activities and experiments, we are thankful to Ted Brattstrom for *Reckoning Latitude*, and Forest Luke for *Tracking Mars*. Both Ted and Forrest are high school teachers and amateur astronomers in Hawaii—Ted at Pearl City High School, and Forrest at Leilehua High School. For valuable feedback and advice on most of the astronomy material, we are grateful to Richard Crowe, University of Hawaii at Hilo, and to John Hubisz, North Carolina State University.

For general suggestions on all aspects of this manual, we remain indebted to mentor Charlie Spiegel. For feedback, we are grateful to Marshall Ellenstein.

Master List of Equipment and Supplies

Item	Activities and Experiments
9-volt batteries	*Sensing pH*
acetate strips	*A Force to be Reckoned*
acetic acid	*Smells Great!*
alcohol, rubbing	*Mystery Powders, Circular Rainbows*
aluminum foil	*I'm Melting! I'm Melting!, Pinhole Camera, Solar Power I*
ammeters, DC (0-5A)	*An Open and Short Case*
ammonia cleanser	*Sensing pH*
aquarium tank or sink	*Sink or Swim*
baking soda	*Mystery Powders, Foamy Bubble Round-Up, Sensing pH*
balance, equal arm	*Spiked Water*
balances (precision)	*Making Cents, Upset Stomach, Sink or Swim, Eureka!, Boat Float*
balls, steel	*Rolling Stop*
bar magnets	*A Force to be Reckoned, Magnetic Personality*
BB shot	*Thickness of a BB Pancake*
beakers	*Mystery Powders, Circular Rainbows*
beakers, 1000-mL	*Foamy Bubble Round-Up*
beakers, 250-mL	*Tubular Rust*
beakers, 600-mL	*Boat Float*
benzyl alcohol	*Smells Great!*
borax	*Sensing pH*
boric acid	*Mystery Powders*
bottles, glass ketchup	*Pure Sweetness*
bowls	*Sink or Swim*
bricks	*A Force to be Reckoned*
brown sugar	*Pure Sweetness*
buckets, 3-gallon	*Boat Float*
bulb sockets, miniature	*Batteries and Bulbs, An Open and Short Case, Be the Battery, Three-Way Switch*
bulbs, miniature	*Batteries and Bulbs, An Open and Short Case, Be the Battery, Three-Way Switch*
buret stands with clamps	*Upset Stomach*
candles	*Tuning the Senses*
cans, radiation	*Canned Heat I, Canned Heat II*
cards, 3" x 5"	*Pinhole Image, Sunballs*
carpet strips (4 m x 50 cm)	*Rolling Stop*
chunks of wood	*Boat Float*
clay, modeling	*Boat Float*
club soda	*Sensing pH*
compasses	*Over and Under, Tracking Mars*
cooking pot	*Pure Sweetness, Sensing pH*
copper acetate	*Crystal Growth*
cornstarch	*Mystery Powders*
countertop defrosting plates	*I'm Melting! I'm Melting!*
crucibles	*Circular Rainbows*
cups, clear plastic	*Sensing pH*
D-cell batteries	*Batteries and Bulbs, An Open and Short Case, Three-Way Switch*
diethyl acetate (finger nail polish remover)	*Circular Rainbows*
dollar bills	*Reaction Time*
Dominoes	*Chain Reaction*

drinking straws	*Reckoning Latitude*
eggs, raw	*Egg Toss, Sink or Swim*
epsom salt	*Mystery Powders*
Erlenmeyer flasks, 250-mL	*Foamy Bubble Round-Up, Upset Stomach*
ethanol	*Oleic Acid Pancake, Circular Rainbows, Name that Recyclable*
eyedropper	*Oleic Acid Pancake, Mystery Powders*
film canisters	*Eureka!*
fishing weights	*Boat Float*
food coloring	*Dance of the Molecules, Solar Power II*
forceps	*Crystal Growth*
garbage bags	*Egg Toss*
Genecon® hand generators	*Be the Battery, Three-Way Switch*
glass slides	*Crystal Growth*
glassine paper	*Pinhole Camera*
gloves	*Pure Sweetness, Upset Stomach*
graduated cylinders, 100-mL	*Thickness of a BB Pancake, Solar Power II*
graduated cylinders, 10-mL	*Oleic Acid Pancake*
graduated cylinders, 500 mL	*Eureka!, Boat Float*
graduated cylinders, wide-mouth	*Sink or Swim*
grapefruit juice	*Sensing pH*
hardness sets	*What's That Mineral?*
heat lamps and bases	*Canned Heat I*
hot plates	*Crystal Growth*
hydrochloric acid (HCl)	*What's That Mineral?, What's That Rock?*
ice cubes	*I'm Melting! I'm Melting!*
iron filings	*Magnetic Personality*
isoamyl alcohol	*Smells Great!*
isobutanol	*Smells Great!*
isopentenol	*Smells Great!*
jars, baby food	*Dance of the Molecules, Pure Sweetness*
jars, glass	*Solar Power I, Indoor Clouds*
kitchen knifes	*Pure Sweetness*
lead blocks	*Sink or Swim*
lens, 25 mm converging	*Pinhole Camera*
light bulbs, clear 100-watt	*Solar Power I*
lycopodium powder	*Oleic Acid Pancake*
magnetic field projectuals	*Magnetic Personality*
masses, hooked	*Bouncy Board, Boat Float*
matches	*Tuning the Senses*
measuring cups	*Indoor Clouds*
metersticks	*Go! Go! Go!, Bouncy Board, Rolling Stop, Pinhole Image, Solar Power I, Solar Power II, Sunballs*
methanol	*Circular Rainbows, Smells Great!*
micrometer	*Thickness of a BB Pancake*
microscopes	*Crystal Growth*
mineral collections	*What's That Mineral?*
mirror, full length	*Mirror, Mirror, on the Wall*
molecular modeling kits	*Molecules by Acme*
mortar and pestle	*Upset Stomach*
nails (short)	*Spiked Water*
n-propanol	*Smells Great!*
nuts, 1/2-inch	*Sugar Soft*
octanol	*Smells Great!*
oleic acid	*Oleic Acid Pancake*
paint, flat black	*Solar Power I*
paper towels	*Spiked Water, Canned Heat I, Canned Heat II, I'm Melting! I'm Melting!*

paper, butcher	*Go! Go! Go!*
paper, circular filter	*Circular Rainbows*
paper, graph	*Making Cents, Go! Go! Go!, Canned Heat I, Canned Heat II, Sugar Soft, Tracking Mars, Tire Pressure and 18-Wheelers*
pencils, colored	*Over and Under*
pennies	*Making Cents*
pens, black felt-tip	*I'm Melting! I'm Melting!, Circular Rainbows*
petri dishes	*Crystal Growth*
phenolphthalein	*Mystery Powders*
pins, straight	*Pinhole Image*
pipets, 9-inch plastic	*Sugar Soft, Upset Stomach*
pith balls	*A Force to be Reckoned*
plaster of Paris	*Mystery Powders*
plastic wrap	*Solar Power II*
playing cubes (or painted sugar cubes	*Get a Half-life*
plumb line and bob	*Reckoning Latitude*
potassium aluminum sulfate (alum)	*Crystal Growth*
propionic acid	*Smells Great!*
protractors	*Over and Under, Walking on Water, Reckoning Latitude, Tracking Mars*
pushpins	*Reckoning Latitude*
ramps (2 meters)	*Rolling Stop*
recyclable plastics	*Name that Recyclable*
red cabbage	*Sensing pH*
ring stands	*Foamy Bubble Round-Up, Tubular Rust*
rock samples	*What's That Rock?*
rubber bands	*Tubular Rust, Solar Power II*
rubber stoppers, drilled	*Foamy Bubble Round-Up, Solar Power I*
rulers, centimeter	*Reaction Time, Mirror, Mirror, on the Wall, Thickness of a BB Pancake, Sugar Soft, Tubular Rust, Top This, Walking on Water, Tracking Mars*
safety goggles	*Egg Toss, Mystery Powders, Salt and Sand, Foamy Bubble Round-Up, Pure Sweetness, Sensing pH, Upset Stomach*
salicylic acid	*Smells Great!*
salt, table	*Mystery Powders, Salt and Sand, Sensing pH, Sink or Swim*
sand	*Salt and Sand*
shoeboxes	*Pinhole Camera*
silk cloth squares	*A Force to be Reckoned*
soda-pop in cans	*Sink or Swim*
sodium chloride	*Crystal Growth*
sodium hydroxide	*Mystery Powders*
sodium nitrate	*Crystal Growth*
spatulas	*Mystery Powders*
spoons	*Sink or Swim*
spring scales	*Boat Float*
steel wool	*Tubular Rust*
stereo audio device	*Sound Off*
stopwatches	*Go! Go! Go!, Chain Reaction*
strainers	*Sensing pH*
streak plates	*What's That Mineral?*
string	*Ellipses, Eureka!, Boat Float*
strobe light, variable frequency	*Slow-Motion Wobbler*
Styrofoam blocks	*Sink or Swim*
Styrofoam cups	*Spiked Water, Solar Power II*
Styrofoam plates	*I'm Melting! I'm Melting!*
sugar, table	*Mystery Powders, Sugar Soft*
switch, double pole double throw	*Sound Off*
switches, single pole double-throw	*Three-Way Switch*

tape, clear	*Solar Power I, Reckoning Latitude*
tape, masking	*Go! Go! Go!, Egg Toss, Mirror, Mirror, on the Wall, Pinhole Camera*
test tube clamps	*Tubular Rust*
test tubes (15 cm)	*Mystery Powders, Tubular Rust*
thermometer (Celsius)	*Spiked Water, Canned Heat I, Canned Heat II, Solar Power I, Solar Power II*
thumbtacks	*Ellipses*
thymol	*Crystal Growth*
tincture of iodine	*Mystery Powders*
tire pressure gauge	*Tire Pressure and 18-Wheelers*
toilet bowl cleaner	*Sensing pH*
topographic maps	*Top This*
toy boat (or small cake pan)	*Boat Float*
toy cars, constant velocity	*Go! Go! Go!*
trays	*Thickness of a BB Pancake, Oleic Acid Pancake, Indoor Clouds*
trundle wheel	*Egg Toss*
tubing	*Foamy Bubble Round-Up*
tuning forks, low frequency	*Slow-Motion Wobbler*
twine	*Pulled Over, Bouncy Board*
vinyl strips	*A Force to be Reckoned*
washing soda	*Mystery Powders, Sensing pH*
weigh dishes	*Upset Stomach*
well-plates	*Upset Stomach*
white chalk	*Mystery Powders*
white vinegar	*Mystery Powders, Foamy Bubble Round-Up, Circular Rainbows, Sensing pH, Tubular Rust*
wires, connecting	*Batteries and Bulbs, An Open and Short Case, Be the Battery, Three-Way Switch*
wood blocks	*Sink or Swim, Boat Float*
wool cloth squares	*A Force to be Reckoned*

Table of Contents

Important Notes About Safety

When performing the activities and experiments in this manual, you should always keep the safety of yourself and others in mind. Most safety rules that must be followed involve common sense. For example, if you are ever unsure of a procedure or chemical, ask your teacher who should always be there to help you.

To guide you to safe practices, in this manual you will find several types of icons posted by selected activities. Here are the icons and what they indicate:

 Wear approved safety goggles. Wear goggles when working with a chemical, solution, or when heating substances.

 Wear gloves. Wear gloves when working with chemicals.

 Flame/heat. Keep combustible items, such as paper towels, away from any open flame. Handle hot items with tongs, oven mitts, pot holders or the like. Do not put your hands or face over any boiling liquid. Use only heat-proof glass, and never point a heated test tube or other container at anyone. Turn off the heat source when you are finished with it.

Here are some specific safety rules that should be practiced at all times:

1. Do not eat or drink in the laboratory.

2. Maintain a clean and orderly work space. Clean up spills at once or ask for assistance in doing so.

3. Do not perform unauthorized experiments. First obtain permission from your teacher. It is most important that others know what you are doing and when you are doing it. Think of it this way: If you are doing an experiment alone and something goes terribly wrong, there will be no one there to help you. That's not using your common sense.

4. Do not taste any chemicals or directly breathe any chemical vapors.

5. Check all chemical labels for both name and concentration.

6. Do not grasp recently heated glassware, clamps, or other heated equipment because they remain hot for quite a while.

7. Discard all excess reagents or products in the proper waste containers.

8. If your skin comes in contact with a chemical, rinse under cold water for at least 15 minutes.

9. Do not work with flammable solvents near an open flame.

10. Assume any chemical is hazardous if you are unsure.

CONCEPTUAL PHYSICAL SCIENCE **EXPLORATIONS**	**Activity**

About Science: Observation

Tuning the Senses

Purpose
In this activity, you will tune your senses of sight and sound.

Scientists' original source of information about the universe comes from personal observations. This leads to questioning reasons and causes. A scientist notices something, asks questions, and then tries to answer them. By this definition, can't we all be scientists?

Required Equipment and Supplies
notebook, pen or pencil, candle, matches, and patience

Discussion
Galileo wrote, "In questions of science the authority of a thousand is not worth more than the humble observation and reasoning of a single individual." We'll do two simple activities, the first to tune our hearing, the second to tune our seeing.

Procedure
Perform the following activities on your own and then answer the questions on the following page.

Activity 1: Audition of the Environment
Go outside and find a comfortable place to sit. Listen to your environment for 10 minutes. Write down the sounds you hear. You might find it helpful to close your eyes during this activity—don't let the sights of your surroundings distract you from observing its sounds.

Activity 2: Observation of a Burning Candle
Remain absolutely quiet while observing an unlit candle for two minutes. Record your observations.

Light the candle, observe it for two minutes, then record your observations.

After you have exhausted all observations possible, extinguish the candle.

Observe the extinguished candle while recording more observations.

What other activities can you invent to tune your senses?

Be creative and tune your senses every day!

Questions for Activity 1: Audition of the Environment

1. What was the quietest sound you heard? What was the loudest?

_____

2. Which sounds had a relatively high pitch (a light breeze, for example)?

_____

3. Which sounds had a relatively low pitch (a car engine, for example)?

_____

4. How many sounds were natural and how many were man-made?

_____

5. Did you identify sounds by their sources or did you spell them out phonetically?

_____

6. Are there any continuous sounds *right now* (like that of an air conditioner) that you have simply tuned out? What are they?

_____

7. How might you describe a sound to someone who was unfamiliar with its source (for example, the sound of a car to someone who had never heard of a car before)?

_____

Questions for Activity 2: Observation of a Burning Candle

8. Did you detect any odors? If so, describe them.

_____

9. What color was the molten wax?

_____

10. Did you note the time and date of your candle observations?

_____

11. Was it hotter six inches above the flame or six inches to the side of the flame (or was it equally hot in both places)?

_____

12. Describe the patterns in the smoke before and after the flame was extinguished.

_____

13. Was the color transition of blue to yellow in the flame gradual or abrupt?

_____

14. Did you describe the candle by drawing a picture of it?

_____

2

CONCEPTUAL PHYSICAL SCIENCE **EXPLORATIONS**	**Activity**

About Science: The Scientific Method

Making Cents

Purpose
In this activity, you will investigate the relationship between the mass of a penny and its age.

Required Equipment and Supplies
10 pennies per student
balances
graph paper

Discussion
The scientific method is an effective way of gaining, organizing, and applying new knowledge. The method is essentially as follows:
1. Recognize a problem.
2. Make an educated guess—a **hypothesis**.
3. Predict the consequences of the hypothesis.
4. Perform experiments to test predictions. If necessary, modify the hypothesis in light of experimental results. Perform more experiments.
5. Formulate the simplest general rule that organizes the three main ingredients—hypothesis, prediction, and experimental outcome.

Procedure
Step 1: Propose a hypothesis to the following question (problem): What effect does aging have on a penny's mass?

Step 2: Based on your hypothesis, predict the general form of a graph that plots the mass of a penny (*y*-coordinate) relative to its age (*x*-coordinate).

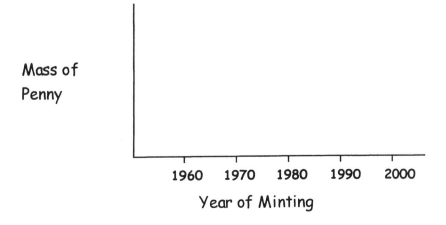

Step 3: Using a balance, measure the mass of at least 10 individual pennies minted in different years. Enter the mass in grams of each penny relative to the year it was minted (Table 1).

Table 1.

Mass of Penny:									
Year Minted:									

Step 4: Pool your data with that of all other students on the class chalkboard, and create your own graph using all this data. Show the mass of each penny in grams on the *y*-coordinate and the year the penny was minted on the *x*-coordinate. Alternatively, data may be entered into a computer program that will plot the graph for you.

Step 5: Are the experimental results consistent with your hypothesis? If not, propose a new hypothesis.

Step 6: If you have formed a new hypothesis, what additional measurements might you take to support this new hypothesis? Perform these measurements and record your results and observations here:

Summing Up

1. What conclusion can you draw from the results of your experimental data?

_____

_____

2. What effect might aging have on the mass of a nickel, a dime, and a quarter?

_____

_____

3. Would using a balance that was many times more sensitive have made a difference in your conclusion about the effect of aging on a penny? Briefly explain.

_____

_____

4. What improvements might you expect in your graph if only one student had done all the weighing on a single balance?

_____

_____

| CONCEPTUAL PHYSICAL SCIENCE **EXPLORATIONS** | **Experiment** |

Newton's First Law: Descriptions of Motion

Go! Go! Go!

Purpose
In this experiment, you will plot a graph that represents the motion of an object.

Required Equipment and Supplies
constant velocity toy car
butcher paper or continuous (unperforated) paper towel
access to tape
stopwatch
meterstick
graph paper

Discussion
Sometimes two quantities are related to each other, and the relationship is easy to see. Sometimes the relationship is harder to see. In either case, a graph of the two quantities often reveals the nature of the relationship. In this experiment, we will plot a graph that represents the motion of a real object.

Procedure
We are going to observe the motion of the toy car. By keeping track of its position relative to time, we will be able to make a graph representing its motion. To do this, we will let the car run along a length of butcher paper. At one-second intervals, we will mark the position of the car. This will result in several ordered pairs of data—positions at corresponding times. We can then plot these ordered pairs to make a graph representing the motion of the car.

Step 1: Fasten the butcher paper to the top of your table. It should be as flat as possible—no hills or ripples.

Step 2: If the speed of the toy car is adjustable, set it to the slow setting.

Step 3: Aim the car so that it will run the length of your table. Turn it on and give it a few trial runs to check the alignment.

Step 4: Practice using the stopwatch. For this experiment, the stopwatch operator needs to call out something like, "Go!" at each one-second interval. Try it to get a sense of the one-second rhythm.

Step 5: Practice the task.
 a. Let the car drive across the length of the butcher paper.
 b. Soon after it starts, the stopwatch operator will start the stopwatch and say, "Go!"
 c. Another person in the group should practice marking the location of the front or back of the car on the butcher paper every time the watch operator says, "Go!" For the practice run, simply touch the eraser of the pencil to the butcher paper at the appropriate points.
 d. The watch operator continues to call out, "Go!" once each second and the marker continues to practice marking the location of the car until the car reaches the end of the butcher paper or table. Take care to keep the car from running off the table!

Step 6: Perform the task.

    a. Let the car drive across the length of the butcher paper.

    b. Soon after it starts, the stopwatch operator will start the stopwatch and say, "Go!"

    c. Another person in the group will mark the location of the front or back of the car on the butcher paper every time the watch operator says, "Go!"

    d. The watch operator continues to call out, "Go!" once each second and the marker continues to mark the location of the car until the car reaches the end of the butcher paper or table. Take care to keep the car from running off the table!

Step 7: Label the marked points. The first mark is labeled "0," the second is labeled "1," the third is "2," and so on. These labels represent the time at which the mark was made.

Step 8: Measure the distances—in centimeters—of each point from the point labeled "0." (The "0" point is 0 cm from itself.) Record the distances on the data table. Don't worry if you don't have as many data points as are spaces available on the data table.

Data Table

Time t (s)	0	1	2	3	4				
Distance d (cm)	0								

Step 9: Make a plot of distance vs. time on the graph paper. Title the graph "Distance vs. Time." Make the horizontal axis time and the vertical axis distance. Label the horizontal axis with the quantity's symbol and the units of measure: "t (s)". Label the vertical axis in a similar manner. Make a scale on both axes starting at 0 and extending far enough so that all your data will fit within the graph. Don't necessarily make each square equal to 1 second or 1 centimeter. Make the scale so the data will fill the maximum area of the graph.

We could just as easily make a graph of time vs. distance. But we prefer distance vs. time for a few reasons. In this experiment, time is what we call an "independent variable." That is, no matter how fast or slow our car was, we always marked its position at equal time intervals. We were in charge of the time intervals; the car was "in charge" of the distance it moved in each interval. But the distance the car moved in each interval depended on the time interval we chose. So we call distance the "dependent variable." We generally arrange a graph so that the horizontal axis represents the independent variable and the vertical axis represents the dependent variable. Furthermore, the slope of a distance vs. time graph is more informative than the slope of a time vs. distance, as we will see later.

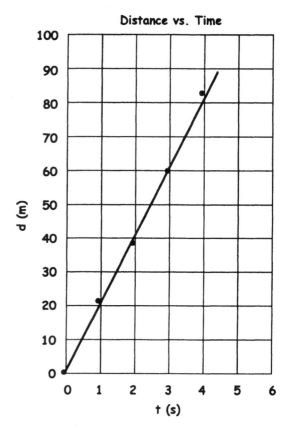

Figure 1.

Step 10: Draw a line of best fit. In this case, the line of best fit should be a single, straight line. Use a ruler or straight edge; place it across your data points so that your line will pass as close as possible to all your data points. The line may pass above some points and below others. Don't simply draw a line connecting the first point to the last point. An example is shown in figure 1.

Step 11: Determine the slope of the line. Slope is often referred to as "rise over run." To determine the slope of your line, proceed as follows.

a. Pick two convenient points on your line. They should be pretty far from each other. Convenient points are those that intersect grid lines on the graph paper.

b. Extend a horizontal line to the right of the lower convenient point, and extend a vertical line downward from the upper convenient point until you have a triangle as shown in figure 2. It will be a right triangle, since the horizontal and vertical lines meet at a right angle.

c. Find the length of the horizontal line on your graph. This is the "run." Don't use a ruler; the length must be expressed in units of the quantity on the horizontal axis. In this case, seconds of time.

Run: _____ s

d. Measure the length of the vertical line. This is the "rise." Don't use a ruler; the length must be expressed in units of the quantity on the vertical axis. In this case, centimeters of distance.

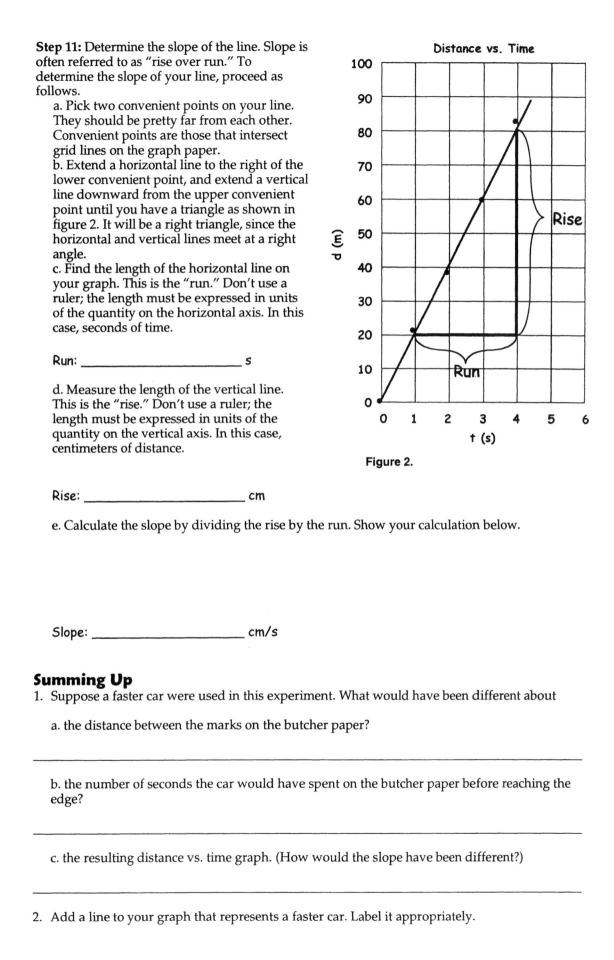

Figure 2.

Rise: _____ cm

e. Calculate the slope by dividing the rise by the run. Show your calculation below.

Slope: _____ cm/s

Summing Up

1. Suppose a faster car were used in this experiment. What would have been different about

a. the distance between the marks on the butcher paper?

_____

b. the number of seconds the car would have spent on the butcher paper before reaching the edge?

_____

c. the resulting distance vs. time graph. (How would the slope have been different?)

_____

2. Add a line to your graph that represents a faster car. Label it appropriately.

3. Suppose a slower car were used in this experiment. What would have been different about

   a. the distance between the marks on the butcher paper?

_____

   b. the number of seconds the car would have spent on the butcher paper before reaching the edge?

_____

   b. the resulting distance vs. time graph. (How would the slope have been different?)

_____

4. Add a line to your graph that represents a slower car. Label it appropriately.

5. Suppose the car's battery ran out during the run so that the car slowly came to a stop.

   a. What would happen to the space between marks as the car slowed down?

_____

   b. Add a line to your graph that represents a car slowing down. Label it appropriately.

6. What motions do these graphs represent? In other words, what was the car doing to generate these motion graphs?

Line A. _____

_____

_____

_____

Line B. _____

_____

_____

_____

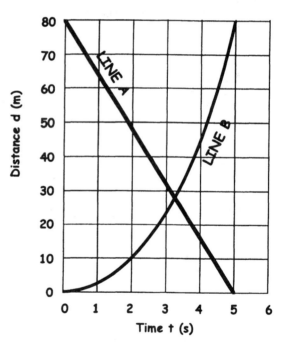

Name_____ Period_____ Date_____

CONCEPTUAL PHYSICAL SCIENCE **EXPLORATIONS**	**Activity**

Newton's First Law: Descriptions of Motion

Sonic Ranger

Purpose

In this activity, you will use graphs to investigate motion. The graphs will represent your own motion and will be drawn by the computer as you move.

Required Equipment and Supplies

sonic ranging device with appropriate equipment and software
computer

Discussion

Graphs can be used to represent motion. For example, if you track the position of an object as time goes by, you can make a plot of position vs. time. In this activity, the sonic ranger will track your position and the computer will draw a position vs. time graph of your motion. The sonic ranger sends out a pulse of high frequency sound and then listens for the echo. By keeping track of how much time goes by between each pulse and corresponding echo, the ranger determines how far you are from it. (Bats use this technique to navigate in the dark.) By continually sending pulses and listening for echoes, the sonic ranger tracks your position over a period of time. This information is fed to the computer, and the software generates a position vs. time graph.

Procedure

Your instructor will provide a computer with a sonic ranging program installed. Check to see that the sonic ranger is properly connected and operating reliably. Position the sonic ranger so that its beam is about chest high and aimed horizontally. (Note: Sometimes these devices do not operate reliably on top of computer monitors.)

The sonic ranger should be set to "long range" mode. The computer should be set to graph position vs. time. Initiate the sonic ranger and note how close and how far you can get before the readings become unreliable.

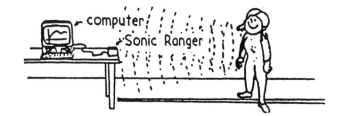

Part A: Move to Match the Graph

Generate real time graphs of each motion depicted below and write a description of each. *Do not use the term "acceleration" in any of your descriptions.* Instead, use terms and phrases such as, "rest," "constant speed," "speed up," "slow down," "toward the sensor," and "away from the sensor."

Study each graph below. When you are ready, initiate the sonic ranger and move so that your motion generates a similar graph. Then describe the motion in words.

Example:

Position vs. Time Graph

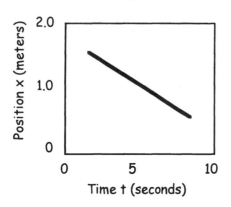

Description

Move toward the sensor at constant speed.

Make sure each person in the group can move to match this graph before moving on to the next graph.

1.

Position vs. Time Graph

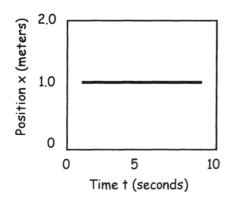

Description

2.

Position vs. Time Graph

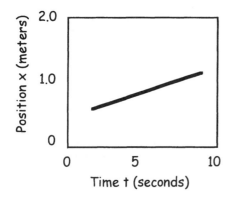

Description

3.

Position vs. Time Graph   **Description**

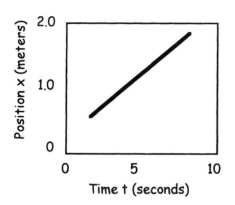

4.

Position vs. Time Graph   **Description**

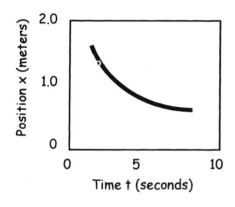

Part B: Move to Match the Words
Walk to match each description of motion. Draw the resulting position vs. time graph.

5.   **Description**   **Position vs. Time Graph**

Move toward the sensor at
constant speed, stop and remain
still for a second, then walk
away from the sensor with
constant speed.

6.   **Description**   **Position vs. Time Graph**

Move toward the sensor with
decreasing speed, then just as
you come to rest, move away
from the detector with
increasing speed.

11

Move away from the sensor with
decreasing speed until you come
to a stop. Then move toward the
sensor with decreasing speed
until you come to a stop.

Summing Up

1. How does the graph show the difference between forward motion and backward motion?

_____

2. How does the graph show the difference between slow motion and fast motion?

_____

3. Study the graph of position vs. time shown below.

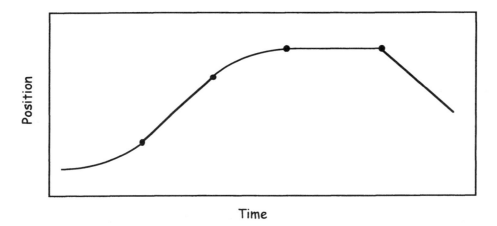

Label sections of the graph showing where the object is
• at rest
• moving forward at constant speed
• moving backward at constant speed
• speeding up
• slowing down

Name_____ Period_____ Date_____

| CONCEPTUAL PHYSICAL SCIENCE **EXPLORATIONS** | **Activity** |

Newton's Second Law: Force, Mass, and Acceleration
Pulled Over

Purpose
In this activity, you will observe the effects of force and inertia.

Required Equipment and Supplies
about 6 feet of twine
table
3 textbooks, such as *Conceptual Physical Science—Explorations* (or 3 equivalent masses)

Discussion
When a net force is applied to an object, its motion changes. It *accelerates*. Gravitational force between an object and the earth (the object's weight) provides a convenient virtually constant force. Drop a book and the interplay of gravity and the book's mass gives it an acceleration g (nearly 10 m/s^2). If the dropping book is attached by a length of twine to another book that lies on a horizontal surface, the acceleration of the dropping book is less—it is restrained by the extra mass of the book above, and friction between the book above and the supporting surface.

Procedure
Place a book, call it Book A, on a horizontal surface. It doesn't accelerate because the downward gravitational force acting on it is balanced by an upward support force (called normal force). With no net force in the vertical direction, there is no vertical acceleration. But if a length of twine is attached to another book that falls vertically, the falling book—call it Book B—provides a force that accelerates both books. Tie two books together by a 3 or 4-foot length of twine, and place Book A in the middle of a table while holding Book B slightly off the edge of the table. Let go, and the books accelerate—Book B vertically, and Book A horizontally. Because they're attached, both have the same amount of acceleration. You'll want to make the connecting twine between the books slightly longer than the height of  the table, so that Book A can coast to the edge after Book B hits the floor. (If you're worried about damaging your textbooks, you should substitute other more durable masses for this activity.)

Make an estimated judgment ("guesstimate") about the amount of acceleration the two-book system undergoes compared to that of free fall, g. Then try it and see.

The string connecting the books exerts a force on Book A. But there is another force at work here, and that force is friction. Friction acts in a direction opposite the motion. If friction is too great, the system doesn't even move when you release Book B. Do what you can to minimize friction on Book A, like using rollers of some kind (round pencils, perhaps). If you mount Book A on a cart, be sure to attach same amount of mass to Book B!

(If you use masses other than books; the main idea is to first try equal masses for A and B, and to minimize friction between A and the surface and between the string and the edge of the table.)

13

Going Further

1. Attach another book to Book A, doubling its weight (*and doubling the friction between it and the table also*). You now have the force provided by the weight of one book acting on itself and two other books—three times the mass. Here's where it's important that you've minimized friction! Predict the fraction of *g* that the books will experience when you release Book B.

2. Repeat, but this time attach the extra book to falling Book B. Now you have twice the force acting on three times the mass (and less friction compared to the previous trial!). Predict the amount of acceleration, compared to the previous cases, and then try it and see.

Summing Up

1. Rank the amount of accelerations for the three cases:

   Greatest acceleration _____

   Mid acceleration_____

   Least acceleration _____

2. When a book is in free fall, the net force on it is its weight. That weight acts on the *mass* of the single book—and the book accelerates at *g*. If you drop two books attached by a length of string, twice as much weight acts on twice as much mass—and it accelerates at

   _____

3. When one of the books is dropped while connected by twine to the other book atop a horizontal table, both books accelerate. Why is the acceleration less than *g*?

   _____

4. When two books were dropped while connected by twine to a single book atop the table, the acceleration was greater than for one book dropping. If three books were dropped while attached to a single book atop the table, would the acceleration be even greater?

   _____

5. Apparently, the greater the load that drops compared to the load pulled horizontally across the tabletop, the greater the acceleration. What is the upper limit of acceleration that can be produced by this method? (Use exaggeration: Suppose the dropped load was huge and the sliding mass very small.)

   _____

6. How much did friction affect your results? How could you better reduce friction?

   _____

14

Momentum: Impulse

Egg Toss

Purpose
In this activity, you will investigate the effect that stopping time has on stopping force when momentum changes.

Required Equipment and Supplies
raw egg
garbage bag (maybe two)
masking tape
safety glasses or goggles
a playing field (soccer, football, baseball, softball, etc.)
trundle wheel (optional)

Procedure
One person in the group is going to throw a raw egg to another person in the group. The second person must catch the egg without letting it break. When the thrower and catcher are close to each other, the task is fairly simple. As the distance increases, the task becomes more difficult.

Step 1: Prepare for the activity by choosing a thrower and a catcher. The thrower will throw the egg to the catcher. The catcher will catch the egg. Both thrower and catcher must use only their bare hands to handle the egg.

Step 2: The catcher must wear the safety goggles and a plastic poncho. Use the garbage bag and masking tape to construct a plastic poncho. When the egg breaks, it can make a mess. Make sure the poncho covers any part of your clothes that should be protected from raw egg white and yolk.

Step 3: Go to the field and line up according to your teacher's instructions. The throwers and catchers should be facing each other and should start about three meters apart from each other. Throwers should also be two to three meters from each other.

Step 4: When the teacher gives the signal, the thrower throws the egg to the catcher. If the catcher catches the egg and the egg remains intact, the group may proceed to the next toss. If the egg breaks, the catcher must remove the poncho and use it to clean up the egg. If a trundle wheel is available, use it to determine the distance between the thrower and catcher when the egg broke.

The thrower (or someone else in the group) must retrieve the egg from the catcher. The catchers should move back three meters, and the throwers should return to their original throwing line.

Step 5: Repeat step 4 until the last group breaks their egg.

Summing Up

1. How far did your thrower and catcher get from each other before the egg broke? What was the longest distance achieved in the class? (Use an estimate if you didn't measure it.)

_____

_____

2. What was the trick to making a successful catch? What does this have to do with stopping time?

_____

_____

3. Compare a sudden-stop catch with a gradual-stop catch.
   a. In which case is the mass of the egg greater? Or is it the same either way?

_____

   b. In which case is the change in velocity of the egg greater? Or is it the same either way? (Be careful!)

_____

   c. In which case is the change in momentum ($m\Delta v$) of the egg greater? Or is it the same either way?

_____

   d. In which case is the stopping time greater? Or is it the same either way?

_____

   e. In which case is the stopping force greater? Or is it the same either way?

_____

4. Use your findings from this activity to explain the purpose of airbags in cars. Don't use words like "cushion," or "absorb." Do use terms like "stopping time," and "stopping force."

_____

_____

5. What are some other examples of changing stopping time to change stopping force?

_____

_____

Momentum: Impulse

Bouncy Board

Purpose
In this activity, you will investigate the effect of stopping time when momentum changes.

Required Equipment and Supplies
table
meterstick
short length (30 to 40 cm) of weak string (ex: mailing parcel twine)
various masses

Discussion
In bungee jumping, it is important that the cord supporting the jumper stretches. If the cord has no stretch, then when full extended it either brings the jumper to a sudden halt or the cord snaps. In either case, ouch! Whenever a falling object is brought to a halt, the force that slows the fall depends on the time it acts. We will see evidence of this in this activity.

Procedure
Through the small hole at the end of a meter stick, thread a piece of string about 30 cm long and tie it in place. (If there is no hole, consider drilling one, or tying the string very tightly. Attach a mass to the other end of the string—try a kilogram for starters. Place the stick on a table top and slide the stick so that most of its length extends over the edge of the table, Figure A. While holding the stick firm to the table, hold the mass slightly beyond the edge of the stick—then drop the mass. The string and the bending of the stick should stop its fall. The string shouldn't break (if it does break, try a stronger string or a smaller mass— experiment). Now repeat this activity, but with only a small portion of the stick extending over the table's edge. What happens now?

Fig. A

Fig. B

Try different configurations to see what conditions result in string breaking. Experiment with different masses, different amounts of meter stick overhang, and different string lengths.

Summing Up

1. Why is it important that a bungee jumper be brought to a halt gradually?

_____

_____

2. How does the *impulse = change in momentum* formula, $Ft = m\Delta v$, apply to this activity?

_____

_____

3. Exactly why did the string break when there was less "give" to the meter stick?

_____

_____

4. The breaking strength of the string most certainly plays a role in this activity. How does the length of the string play a role?

_____

_____

5. How does the falling mass play a role? Will twice the mass require twice the stopping force if it is brought to a halt in the same time? Defend your answer.

_____

_____

6. Why is it important that fishing rods bend?

_____

_____

CONCEPTUAL PHYSICAL SCIENCE **EXPLORATIONS**	**Experiment**

Energy: Transformations

Rolling Stop

Purpose
In this experiment, you will investigate the relationship between the height of a ball rolling down an incline and its stopping distance when it rolls off the incline.

Required Equipment and Supplies
about a 2-meter ramp (ex: 5/8 inch aluminum channel) with support about 1/2 meter high
steel ball
meterstick
carpet floor, or piece of *flat* carpet strip (about 4 m long and about 50 cm wide)

Discussion
Mechanical energy is the product of force and distance. When we exert a force to change the energy of an object, we do *work* on an object. A measure of that work, or change in energy, is force × distance. Elevate a ball and we do work on it. With a force equal to its weight we lift it a certain height against gravity. The work done gives it *energy of position*—gravitational potential energy. Raise it twice as high and it has twice the energy. Another way of saying it has twice the energy is to say it has twice the ability to *do* work. When it rolls to the bottom of the incline it can do work on whatever it interacts with. If it has twice the energy, it can do twice the work. An easier-to-visualize example is that of a crate sliding onto a factory floor. In sliding, it does work on the floor and heat it up as it skids. This work is the force of friction × distance of sliding. The question is raised: Will it skid twice as far if it has twice the energy? We'll answer this question not by sliding crates down a ramp (for much of their potential energy would go into heating the ramp itself, which complicates matters), but by rolling balls down inclines and then onto a carpet.

Procedure
Step 1: Mark the ramp at 30 cm, 60 cm, 90 cm, and 120 cm from the bottom end. Assemble the ramp so that when you roll a ball down it, the length of carpet it rolls onto is sufficient to stop it. Experiment to see how far this is.

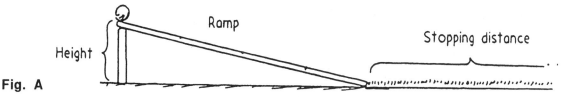

Fig. A

Step 2: Release the steel ball at each of the intervals along the ramp. Measure the vertical height from the floor or table. Roll the ball three times from each height and record the stopping distances in Step 2 Data Table.

Step 3: Change the angle of the ramp, but launch the ball from the *same vertical height* that you did for the previous ramp position.

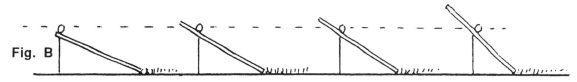

Fig. B

STEP 2 DATA

Initial Position of Ball (cm)	Initial Height of Ball (cm)	Stopping distance (cm)			Average Stopping Distance (cm)
		Trial 1	Trial 2	Trial 3	
30					
60					
90					
120					

STEP 3 DATA

Initial Position of Ball (cm)*	Initial Height of Ball (cm)**	Stopping distance (cm)			Average Stopping Distance (cm)
		Trial 1	Trial 2	Trial 3	

*Because the angle is different, these will no longer be 30cm, 60cm, 90cm, and 120cm.

**These values must be the same as the values in the STEP 2 DATA table.

Going Further: On the graph paper provided by your instructor, construct a graph of average stopping distance (vertical axis) versus height (horizontal axis) for the ball in step 2.

Summing Up

1. Do the graphs indicate direct proportions between height of release and stopping distance?

_____

2. Would you get the same relationship between distance up the ramp (parallel to the ramp rather than height) and stopping distance? Why or why not?

_____

3. How did the stopping distances for the different balls compare? How did the relationship between release heights and stopping distances compare?

_____

4. Nowhere in this activity is there a mention of kinetic energy—energy of motion (KE). A little study of energy conservation tells us that the gain in KE of the ball as it rolls down the ramp is equal to the decrease in PE as the ball loses height—in short, $\Delta PE = \Delta KE$. Why were we able to bypass KE in our analysis here? (The answer to this question underlies the reason that physics types use energy principles to solve problems—intermediate steps can be skipped!)

_____

_____

CONCEPTUAL PHYSICAL SCIENCE **EXPLORATIONS**	**Activity**

Gravity: Free Fall

Reaction Time

Purpose
In this activity, you will measure your personal reaction time.

Required Equipment and Supplies
dollar bill
centimeter ruler

Discussion
Reaction time is the time interval between receiving a signal and acting on it—for example, the time between a tap on the knee and the resulting jerk of the leg. Reaction time often affects the making of measurements. Consider using a stopwatch to measure the time for a 100-meter dash. The watch is started after the gun sounds and is stopped after the tape is broken. Both actions involve reaction time.

Procedure
Step 1: Hold a dollar bill so that the mid-point hangs between your partner's fingers. Challenge your partner to catch it by snapping his or her fingers shut when you release it.

The distance the bill will fall is found using

$$d = \tfrac{1}{2}at^2$$

Simple rearrangement gives the time of fall in seconds.

$$t^2 = \frac{2d}{g}$$

$$t = \sqrt{\tfrac{2}{980}}\sqrt{d}$$

$$t = 0.045\sqrt{d}$$

(For d in centimeters and t in seconds, we use $g = 980 \text{ cm/s}^2$.)

Step 2: Have your partner similarly drop a centimeter ruler between your fingers. Catch it and note the number of centimeters that passed during your reaction time. Then calculate your reaction time using the formula

$$t = 0.045\sqrt{d}$$

where d is the distance in centimeters.

Reaction time = _____

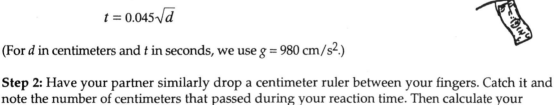

Summing Up

1. What is your evidence for believing or disbelieving that your reaction time is always the same? Is your reaction time different for different stimuli?

_____

_____

2. Suggest possible explanations why reaction times are different for different people.

_____

_____

3. When might reaction time significantly affect measurements you might make using instruments for this course? How could you minimize its role?

_____

_____

4. What role does reaction time play in applying the brakes to a car in an emergency situation? Estimate the distance a car travels at 100 km/h during your reaction time in braking.

_____

_____

5. Give examples in which reaction time is important in sports.

_____

_____

Name_____ Period_____ Date_____

Projectile Motion: Horizontal and Vertical Motion

The Big BB Race

Purpose
In this activity, you will compare the path of a projectile launched horizontally with that of an object in free fall.

Required Equipment and Supplies
simultaneous launcher / dropper

Discussion
Suppose a ball bearing (BB) were launched horizontally at the same time another BB were dropped from the same height. Which one would reach the ground first?

Procedure
Step 1: Draw the path you think the launched projectile will take on figure 1. That is, draw a line connecting the launch point and the impact point in the diagram that traces the path you think the BB will follow.

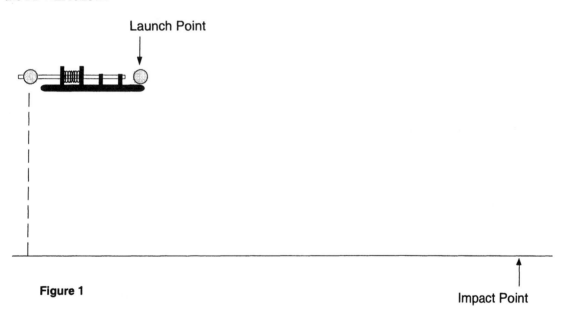

Launch Point

Figure 1

Impact Point

Step 2: Consider the predictions of three students. Write arguments supporting each prediction (whether you agree with the prediction or not).

Student X says the dropped BB will hit first. Explain why this might happen.

_____

_____

Student Y says the fired BB will hit first. Explain why this might happen.

_____

_____

Student Z says the dropped one will hit first. Explain why this might happen.

_____

_____

Step 3: One of the reasons sometimes offered to support the prediction that the dropped ball will hit first is that the launched ball will travel forward for some distance before starting to move downward. What factors might determine the length of this "no-fall distance," as shown in figure 2.

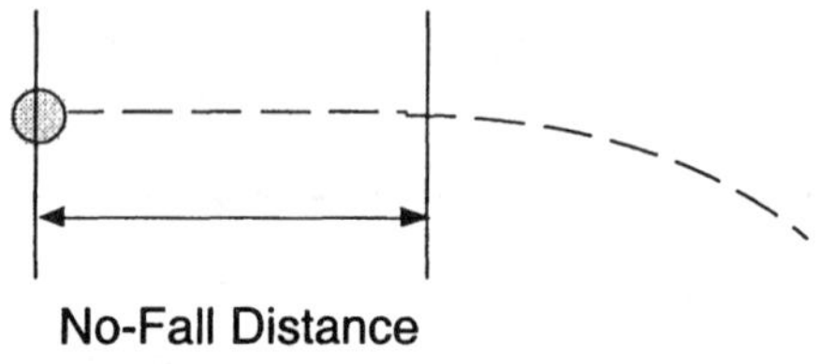

No-Fall Distance

Figure 2

_____

_____

_____

Step 4: Which prediction do you agree with: dropped ball hits first, fired ball hits first, or both hit at the same time?

_____

Step 5: Observe the operation of the simultaneous launch/drop mechanism.
    a. Observe a dropped BB.

    b. Observe a launched BB.

    c. Observe The Big BB Race: a simultaneous launch *and* drop.

Step 6: Which BB hit the ground first, or was it a tie?

_____

Summing Up
1. How does the horizontal motion of a projectile affect the vertical motion of the projectile? In other words, does the horizontal motion of the projectile make it move faster or slower in the vertical direction (or does it have no effect)?

_____

_____

2. Which factors—if any—appear to have the greatest effect on the no-fall distance discussed above?

_____

_____

3. If the launched BB had a rocket engine propelling it forward after it was launched, what—if anything—would have been different about the outcome of The Big BB Race?

_____

_____

Thermal Energy: Temperature

Dance of the Molecules

Purpose
In this activity, you will investigate the difference between hot water and cold water—on the *molecular* level.

Required Equipment and Supplies
source of cold water
source of hot water
two empty baby food jars or small beakers
food coloring

Discussion
The difference between hot things and cold things is referred to as temperature. The temperature of an object depends on the kinetic energy of the random motions of its molecules. A thermometer can be used to measure temperature, but we can't always tell a hot thing from a cold thing just by looking at it. Consider a glass of hot water and a glass of cold water side by side. The water molecules in each glass don't appear to be moving differently from each other. But they are. Follow the steps below to *see* the difference.

Procedure
Step 1: Fill one jar with cold water and the other with hot water.

Step 2: Spend a minute or two making a prediction. What will happen if a few drops of food coloring are added to each jar of water? Record your prediction using words and pictures. (Let the water stand while you record your prediction. This will allow swirling currents and turbulence in the water to diminish.)

_____

_____

_____

Step 3: Carefully add just a drop or two of food coloring to each jar. Watch the coloring spread out for one minute before recording your observations, taking care not to disturb the water in any way. Record your observations using words and pictures.

_____

_____

_____

Summing Up

1. How do your observations support the principle that the molecules in hot objects move faster than the molecules in cold objects?

_____

_____

2. Suppose you let the experiment continue for several hours. What would the jars of water and food coloring look like afterward?

_____

_____

3. Consider a fragrant candle burning in the corner of one side of a large room. It would be possible to smell the fragrance on the other side of the room after a short period of time. Use your observations to explain how this is so.

_____

_____

4. Would the fragrance get to you more quickly, more slowly, or in the same amount of time if the air in the room were colder?

_____

_____

CONCEPTUAL PHYSICAL SCIENCE **EXPLORATIONS**	**Experiment**

Thermal Energy: Specific Heat Capacities

Spiked Water

Purpose
In this experiment, you will determine which is better able to heat a quantity of water: a mass of hot nails or the same mass of equally hot water.

Required Equipment and Supplies
equal arm balance (Harvard trip balance or equivalent)
4 large insulated cups
bundle of short, stubby nails tied together with string
thermometer (Celsius)
hot and cold water
paper towels

Discussion
Suppose you have cold feet when you go to bed, and you want something to keep your feet warm throughout the night. Would you prefer to have a bottle filled with hot water, or one filled with an equal mass of nails at the same temperature as the water? The one that can store more thermal energy will do the better job. But which one is it? In this experiment, we'll find out.

Procedure
1. Which do you think can hold more thermal energy, the nails or the water? Make a prediction before carrying out the experiment.

_____

_____

Step 1: Fill two cups 1/3 full of cold water. Place one cup on each pan of the balance to make sure they contain equal masses of water. Add water to the lighter side until they do. Test to make sure your bundle of nails can be completely submerged in the cold water. If they cannot, remove nails until they can.

Step 2: Dry the nails.

Step 3: Place a large empty cup on each pan of a beam balance. Place the bundle of nails into one of the cups. Add hot water to the other cup until it balances the cup of nails. This should not require more than 1/2 of the cup be filled. When the two cups are balanced, the same mass is in each cup—a mass of nails in one, and an equal mass of water in the other.

Step 4: Lower the bundle of nails into the hot water. Be sure that the nails are completely submerged by the hot water. Allow the nails and the water to reach thermal equilibrium. (This will take a minute or two.)

Step 5: Once the nails and hot water have come to thermal equilibrium, take the nails out of the hot water and put them in one of the cups of cold water. Pour the remaining hot water into the other cup of cold water.

Step 6: Measure and record the final temperature of the mixed water.

$$T_W = \underline{\hspace{4cm}}\,°C$$

Step 7: When the temperature of the nail-water mix stops rising, measure and record it.

$$T_N = \underline{\hspace{4cm}}\,°C$$

Summing Up

1. Which was hotter before being put into the cold water, the nails or the hot water the nails were soaking in? How do you know?

   _____

   _____

2. Which was more effective in raising the temperature of the cold water, the hot nails or the hot water?

   _____

   _____

3. Suppose you have cold feet when you go to bed, and you want something to warm your feet throughout the night. Would you prefer to have a bottle filled with hot water, or one filled with an equal mass of nails at the same temperature as the water? Explain.

   _____

   _____

4. A student who conducted this experiment suggested that the temperature of the nail-water mix rose more than it should have because some water clung to the nails when they were transferred from the hot water to the cold water. Another student says that this caused the temperature of the nail-water mix to rise less than it should have if no water clung to the nails during the transfer. Who do you believe and why?

   _____

   _____

   _____

   _____

CONCEPTUAL PHYSICAL SCIENCE **EXPLORATIONS**	**Experiment**

Heat Transfer and Change of Phase: Absorption

Canned Heat I

Purpose

In this experiment, you will compare the ability of different surfaces to absorb thermal radiation.

Required Equipment and Supplies

heat lamp and base
radiation cans: silver, black, and white
thermometer
access to cold tap water
paper towel
graph paper

Discussion

Does the color of a surface make a difference in how well it absorbs thermal radiation? If so, how? The answer to these questions could help you decide what to wear on a hot, sunny day and what color to paint your house if you live in a hot, sunny climate. In this experiment, we will compare the thermal absorption ability of three surfaces: silver, black, and white. We'll do this by filling cans with these surfaces with water, then exposing the cans to heat lamps. We'll measure the temperature of the water in each of the cans while they're being exposed to the heat and see if there's a difference in the rate at which the temperatures increase.

Procedure

1. In which of the three cans do you think the water will heat up at the fastest rate?

_____

2. In which of the three cans do you think the water will heat up at the slowest rate?

_____

Step 1: Arrange the apparatus so that the heat lamp will shine equally on all three cans. The cans should be about one foot in front of the lamp. Do not turn the light on yet.

Step 2: Fill the cans with cold water and wipe up any spills. Quickly measure the initial temperature of the water in each of the cans and record it in Table A.

Step 3: Turn on the lamp and start timing. Place the thermometer in the silver can.

Step 4: At the 1-minute mark (1:00), read the temperature of the water in the silver can and record it in Table A. Quickly move the thermometer to the black can. Gently swirl the water in the can with the thermometer.

Step 5: At the 2-minute mark (2:00), read the temperature of the water in the black can and record it in Table A. Quickly move the thermometer to the white can. Gently swirl the water in the can with the thermometer.

Step 6: At the 3-minute mark (3:00), read the temperature of the water in the white can and record it in Table A. Quickly move the thermometer to the silver can. Gently swirl the water in the can with the thermometer.

Step 7: Repeat steps 4-6 until the 21-minute mark temperature reading is made.

Table A

Silver Can Temperatures (°C)	Black Can Temperatures (°C)	White Can Temperatures (°C)
Initial	Initial	Initial
T at 1:00	T at 2:00	T at 3:00
T at 4:00	T at 5:00	T at 6:00
T at 7:00	T at 8:00	T at 9:00
T at 10:00	T at 11:00	T at 12:00
T at 13:00	T at 14:00	T at 15:00
T at 16:00	T at 17:00	T at 18:00
T at 19:00	T at 20:00	T at 21:00

Step 8: Turn off the heat lamp.

3. Plot your data for all three cans on a single temperature vs. time graph.

Step 9: Determine the change in temperature for the water in each can while the heat lamp was on.

5. Determine the temperature change in the silver can while the heat lamp was on. Subtract the 1-minute mark temperature reading from the 19-minute mark temperature reading.

   Temperature change in silver can: _____°C

6. Determine the temperature change in the black can while the heat lamp was on. Subtract the 2-minute mark temperature reading from the 20-minute mark temperature reading.

   Temperature change in black can: _____°C

7. Determine the temperature change in the white can while the heat lamp was on. Subtract the 3-minute mark temperature reading from the 21-minute mark temperature reading.

   Temperature change in white can: _____°C

Summing Up

1. In which can did the water heat up at the fastest rate? In which can did the water heat up at the slowest rate? Did your observations match your predictions?

   _____

2. Which would be a better choice if you were going to spend a long time outdoors on a hot, sunny day: a black T-shirt or a white T-shirt?

   _____

3. What happens to the thermal radiation that falls on each of the cans: is it absorbed or reflected?

   Silver:_____ Black:_____ White:_____

32

CONCEPTUAL PHYSICAL SCIENCE EXPLORATIONS | **Experiment**

Heat Transfer and Change of Phase: Thermal Radiation

Canned Heat II

Purpose
In this experiment, you will compare the ability of different surfaces to radiate thermal energy.

Required Equipment and Supplies
radiation cans: silver and black
thermometer
access to hot water
paper towel
graph paper

Discussion
Does the color of a surface make a difference in how well it radiates thermal energy? If so, how? The answer to these questions could help you decide what color coffee pot will best keep its heat and what color might be used to radiate heat away from a computer chip. In this experiment, we will compare the thermal radiation ability of two surfaces: silver and black. We'll do this by filling cans with these surfaces with hot water, then letting the water cool down. We'll measure the temperature of the water in both cans while they're cooling down and see if there's a difference in the rate at which the temperatures decrease.

Procedure
1. In which of the two cans do you think the water will cool down at the faster rate?

_____

2. In which of the two cans do you think the water will cool down at the slower rate?

_____

Step 1: Carefully fill the cans with hot water and wipe up any spills. Quickly measure the initial temperature of the water in each of the cans and record it in Table A. Place the thermometer in the silver can.

Step 2: At the 1-minute mark, read the temperature of the water in the silver can and record it in Table A. Quickly move the thermometer to the black can. Gently swirl the water in the can with the thermometer.

Step 3: At the 2-minute mark, read the temperature of the water in the black can and record it in Table A. Quickly move the thermometer to the white can. Gently swirl the water in the can with the thermometer.

Step 4: Repeat steps 2-3 until the 20-minute mark temperature reading is made.

Table A

Silver Can Temperatures (°C)	Black Can Temperatures (°C)
Initial	Initial
T at 1:00	T at 2:00
T at 3:00	T at 4:00
T at 5:00	T at 6:00
T at 7:00	T at 8:00
T at 9:00	T at 10:00
T at 11:00	T at 12:00
T at 13:00	T at 14:00
T at 15:00	T at 16:00
T at 17:00	T at 18:00
T at 19:00	T at 20:00

3. Plot your data for both cans on a single temperature vs. time graph.

Step 5: Determine the change in temperature for the water in each can while it was allowed to cool.

4. Determine the temperature change in the silver can. Subtract the 1-minute mark temperature reading from the 19-minute mark temperature reading.

   Temperature change in silver can: _____ °C

5. Determine the temperature change in the black can. Subtract the 2-minute mark temperature reading from the 20-minute mark temperature reading.

   Temperature change in black can: _____ °C

Summing Up

1. In which can did the water cool down at the faster rate? In which can did the water cool down at the slower rate? Did your observations match your predictions?

_____

2. What color should the surface of a coffee pot be in order to keep the hot water inside it hot for the longest time?

_____

3. The central processing units (CPUs) in personal computers sometimes get very hot. To prevent damage and improve processing speed, they need to be kept cool. Often, piece of metal is placed in contact with the CPU to draw some heat away. The metal then radiates its heat to the surroundings. To best remove heat from the CPU, should the metal's surface be colored black or left silvery?

Heat Transfer: Conduction and Radiation

I'm Melting! I'm Melting!

Purpose
In this activity, you will investigate curious heat transfer ability of different surfaces.

Required Equipment and Supplies
Miracle Thaw® (or equivalent countertop defrosting plate)
a second Miracle Thaw wrapped tightly in aluminum foil
white Styrofoam plate
black Styrofoam plate (white plate colored completely using a felt-tip pen)
four ice cubes of similar size
paper towel

Discussion
If you walk around the house with bare feet, you probably notice that a tile floor feels much colder than a carpeted floor or rug. It's hard to believe that they might actually have the same temperature. The tile feels colder because it is a better conductor than carpet. Heat is conducted from your warmer feet to the cooler floor faster when the floor is tile than when the floor is carpet. So your feet are cooled faster by tile than they are by carpet at the same cool temperature. In this activity, you will see which kinds of surfaces transfer heat most rapidly.

Procedure
Step 1: Set the Miracle Thaw, Miracle Thaw covered in aluminum foil, white Styrofoam plate, and blackened Styrofoam plate out on your table.

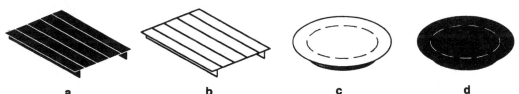

a   b   c   d

**Figure 1. a. Miracle Thaw, b. Miracle Thaw covered in aluminum foil,
c. white Styrofoam plate, d. blackened Styrofoam plate**

1. Which surfaces feel colder and which ones feel warmer? (Just touch a corner; don't transfer too much of your own body heat to any of the objects.)

_____

Step 2: In a moment, you will set an ice cube on each of the surfaces. When you do, they will begin melting. Before you set the ice cubes out, make some predictions.

2. Which ice cube will melt most quickly?

_____

3. Which ice cube will melt most slowly?

_____

Step 3: Set the ice cubes out on their respective surfaces quickly. Observe the ice cubes for several minutes (preferably until the fastest-melting ice cube melts completely).

4. Which ice cube melted most quickly?

_____

5. Which ice cube is melting most slowly?

_____

Summing Up

1. How do your observations compare to your predictions?

_____

_____

2. Which way did heat flow in this activity? (From what to what?)

_____

_____

3. Two of the surfaces were conductors and two of the surfaces were insulators. Which were which?

_____

_____

4. What advantage—in terms of heat transfer—did one Miracle Thaw have over the other?

_____

_____

5. What advantage—in terms of heat transfer—did one Styrofoam plate have over the other?

_____

_____

6. Which of the following conclusions are supported by your observations and which are not? Give evidence from this activity to justify your conclusion.

   a. "Metals transfer heat faster than Styrofoam."

_____

_____

   b. "Black surfaces transfer heat faster than non-black surfaces."

_____

_____

Name_____ Period_____ Date_____

Electricity: Electrostatics

A Force to be Reckoned

Purpose
In this investigation, you will explore the nature of a force. You will determine whether or not the force is distinct from other known forces.

Required Equipment and Supplies
a pith ball suspended from a support rod
vinyl and acetate strips or tubes
wool and silk cloth swatches (2"×2")
brick (or a heavy book)
bar magnet

Procedure
Step 1: Generate an attractive force. Rub the vinyl with the wool. Hold the vinyl near the pith ball and see if the strip will attract the pith ball to it as shown. If it won't, try rubbing the vinyl more vigorously, or try using the acetate rubbed with silk. If you continue to have difficulty, ask your teacher for assistance.

Once you have seen the plastic strip (vinyl or acetate) attract the pith ball, set the plastic aside and discharge the pith ball by gently touching it.

Consider the suggestion that the attraction you observed is simply the result of *gravitational* force. After all, gravitational force causes any two things with mass to be attracted to each other. Recall that the force of attraction is proportional to the amount of mass involved.

Step 2: Hold the brick close to the pith ball. The mass of the brick is much greater than the mass of the plastic strip. What—if anything—does the brick do to the pith ball? What does this tell you about the suggestion that the attraction between the plastic strip and the pith ball is *gravitational?*

_____

_____

Step 3: Generate a repulsive force. Using the plastic strip (rubbed with the cloth as was done in step 1), try to repel the pith ball as shown. Does the observation of repulsion support or contradict the suggestion that the attraction between the plastic strip and the pith ball is gravitational? Why?

_____

_____

Once you have seen the plastic strip (vinyl or acetate) attract the pith ball, set the plastic aside and discharge the pith ball by touching it.

Consider the suggestion that the attraction and repulsion you observed are simply the result of *magnetic* force. After all, magnets can both attract *and* repel.

Step 4: Hold the bar magnet close to the pith ball. What does the magnet do to the pith ball, and what does this tell you about the suggestion that the attraction and repulsion between the plastic strip and the pith ball are magnetic?

_____

_____

Summing Up
The force between the plastic strip and the pith ball is different from gravitational force and different from magnetic force. It is called *electrostatic* force, and is a force between any two objects that are electrically charged.

1. When the vinyl is rubbed with wool, the vinyl gets a *negative* charge. What kind of charge is on the pith ball when the charged vinyl repels it?

_____

2. Does electrostatic force get stronger or weaker with distance? Does the interaction become stronger when charged objects get closer together or when they get farther apart?

_____

_____

Electricity: Electric Circuits

Batteries and Bulbs

Purpose
In this activity, you will explore various arrangements of batteries and bulbs, and the effects of those arrangements on bulb brightness.

Required Equipment and Supplies
2 D-cell batteries
4 connecting wires
2 miniature bulbs (1.5-volt or 2.5-volt flashlight bulbs)
2 miniature bulb sockets

Discussion
Many devices include electronic circuitry, most of which are quite complicated. Complex circuits are made, however, from simple circuits. In this activity you will build one of the simplest yet most useful circuits ever invented—that for lighting a light bulb!

Procedure
Step 1: Remove the bulb from the mini socket.
1. On a separate sheet, draw a detailed diagram of the bulb, showing the following parts of the bulb's "anatomy."
- glass bulb
- screw base
- filament
- filament leads (tiny wires that lead to the filament)
- lead separator (glass bead)
- base insulation ring
- base contact (made of lead)

2. There are four parts of the bulb's anatomy that you can touch (without having to break the bulb). Two of them are made of *conducting* material (metal) and two are made of *insulating* material. List them.

Conducting parts on the outside of the bulb:_____

Insulating parts on the outside of the bulb: _____

Step 2: Examine the two diagrams of a working electric circuit shown below. The diagram on the left shows pictorial representations of circuit elements. The diagram on the right shows symbolic representations. Use the symbolic representations in the steps that follow.

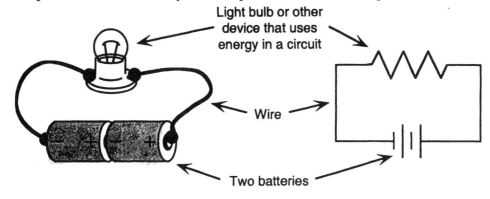

Light bulb or other device that uses energy in a circuit

Wire

Two batteries

Using a bare bulb (out of its socket), one battery, and **two** wires, try lighting the bulb in as many ways as you can. On a separate sheet, sketch at least two *different* arrangements that work. Also sketch at least two arrangements that don't work. Be sure to label them as "works" or "doesn't work."

Step 3: Using a bare bulb (out of its socket), one battery, and **one** wire, try lighting the bulb in as many ways as you can. Sketch your arrangements and note the ones that work.

3. Is it possible to light the bulb using the battery and **no** wires? Explain.

Step 4: Connect one bulb (in its socket) to two batteries as shown in Figure B. This arrangement is often referred to as a *simple* circuit.

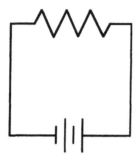

Fig. A Simple circuit

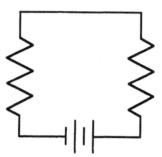

Fig. B Series circuit

Step 5: Connect the bulbs, batteries, and wire as shown in Figure B. When the bulbs are connected one after the other like this, the result is called a *series* circuit.

4. How does the brightness of each bulb in the series circuit compare to the brightness of the bulb in a simple circuit?

---

5. What happens if one of the bulbs in a series circuit is removed? (Do this by unscrewing a bulb from its socket.)

Step 6: Connect the bulbs, batteries, and wire as shown in Figure C. When the bulbs are connected along separate paths like this, the result is called a *parallel* circuit.

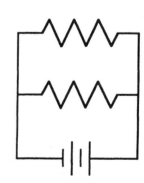

6. How does the brightness of each bulb in the parallel circuit compare to the brightness of the bulb in a simple circuit?

---

7. What happens if one of the bulbs in a parallel circuit is removed?

Fig. C Parallel circuit

Summing Up

1. With what two parts of the bulb does the bulb socket make contact?

_____

_____

2. What do successful arrangements of batteries and bulbs have in common?

_____

_____

3. How do you suppose most of the circuits in your home are wired—in series or in parallel? What is your evidence?

_____

_____

4. How do you suppose automobile headlights are wired—in series or in parallel? What is your evidence?

_____

_____

Name_____ Period_____ Date_____

Electricity: Electric Circuits

An Open and Short Case

Purpose
In this activity, you will explore open circuits and short circuits.

Required Equipment and Supplies
2 D-cell batteries
5 connecting wires
2 miniature bulbs (1.5-volt or 2.5-volt flashlight bulbs)
2 miniature bulb sockets
DC ammeter (0-5A)

Discussion
Electrical circuits are all around us. (We often appreciate them most when the power goes out.)
Most circuits work perfectly well (when power is available). But electrical circuits *can* fail.
Two common modes of circuit failure are *open circuits* and *short circuits*. In this activity, we will
learn how these circuit failures are similar and how they are different.

We will be using an *ammeter* to help us with this investigation. An ammeter is a simple device
used to measure electric current—the rate at which charge flows through a circuit. Current is
measured in amperes ("amps").

Procedure
Part A: Open and Short Circuits
Step 1: Arrange a simple circuit using two batteries, a bulb, an
ammeter, and three connecting wires as shown in figure 1. If the
circuit is working, the bulb will light and some amount of current
will register on the ammeter. It should be less than 1 amp.

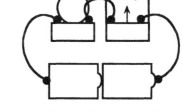

Fig. 1 Simple circuit

Step 2: Predict what would happen to the simple circuit if one of
the wires were disconnected at one point in the circuit. (Don't
touch the circuit yet—predict first!)

1. What will happen to the bulb and what will happen to the
reading on the ammeter (compared to what happened in Step 1)?

_____

_____

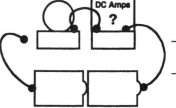

Fig. 2 Open circuit

Once you've made your prediction and discussed it with your
partner(s), disconnect a wire as shown in figure 2. This is an *open
circuit*.

2. Record your observations.

_____

_____

43

Step 3: Predict what would happen to the simple circuit if an additional wire were added to the circuit so as to connect the terminals of the bulb to each other as shown in figure 3. (Don't touch the circuit yet—predict first!)

3. What will happen to the bulb and what will happen to the reading on the ammeter (compared to what happened in Step 1)?

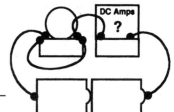

_____

_____

Once you've made your prediction and discussed it with your partner(s), add a wire as shown in figure 3. This is a *short circuit*.

4. Record your observations.

_____

_____

Step 4: One of these circuit failures is said to have almost *no electrical resistance* and one is said to *have infinite electrical resistance*. Electrical resistance is inversely proportional to electrical current in a simple circuit.

5. Which circuit failure has no current and therefore infinite electrical resistance?

_____

6. Which circuit failure has a large amount of current and therefore almost no electrical resistance?

_____

Part B: Short-Circuited Series Circuit

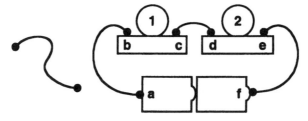

Fig. 4 Series circuit and additional wire

Step 5: Arrange the series circuit shown in figure 4. Notice that bulbs 1 and 2 light up. Notice there is an additional wire not yet in the circuit.

Step 6: Add the additional wire to the circuit, connecting point a to point b. Notice that both bulb 1 and bulb 2 remain fully lit.

Step 7: Now use the additional wire to connect point b to point c. Notice that bulb 1 goes out (or becomes much dimmer) while bulb 2 remains fully lit (or becomes brighter).

7. Predict what will happen if the additional wire is used to connect other points on the circuit. Make your prediction in terms of what will happen to each of the bulbs. Will bulb 1 remain lit or go out? Will bulb 2 remain lit or go out?

Important Note: For purposes of this activity, "remaining lit" includes increased brightness, and "going out" includes significant dimming.

Remember to make predictions before making observations!

a. If point c is connected to point d, bulb 1 will __remain lit __go out and bulb 2 will __remain lit __go out.

b. If point d is connected to point e, _____

c. If point e is connected to point f, _____

d. If point f is connected to point a, _____

e. If point a is connected to point c, _____

f. If point a is connected to point d, _____

g. If point a is connected to point e, _____

h. If point b is connected to point d, _____

i. If point b is connected to point e, _____

Step 8: Observe what happens if the additional wire is used to make each of the connections.

a. When point c is connected to point d, bulb 1 __remains lit __goes out and bulb 2 __remains lit __goes out.

b. When point d is connected to point e, _____

c. When point e is connected to point f, _____

d. When point f is connected to point a, _____

e. When point a is connected to point c, _____

f. When point a is connected to point d, _____

g. When point a is connected to point e, _____

h. When point b is connected to point d, _____

i. When point b is connected to point e, _____

Part C: Short-Circuited Parallel Circuit

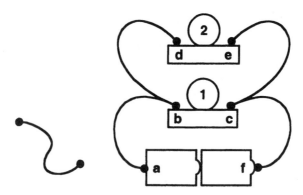

Fig. 5 Parallel circuit and additional wire

Step 9: Arrange the parallel circuit shown in figure 5. Notice that bulbs 1 and 2 light up. Notice there is an additional wire not yet in the circuit.

Step 10: Add the additional wire to the circuit, connecting point a to point b. Notice that both bulb 1 and bulb 2 remain fully lit.

Step 11: Now use the additional wire to connect point b to point c. Notice that bulb 1 goes out (or becomes much dimmer) while bulb 2 remains fully lit (or becomes brighter).

8. Predict what will happen if the additional wire is used to connect other points on the circuit. Make your prediction in terms of what will happen to each of the bulbs. Will bulb 1 remain lit or go out? Will bulb 2 remain lit or go out? **For purposes of this activity, "remaining lit" includes increased brightness, and "going out" includes significant dimming.** Remember to make predictions before making observations!

a. If point c is connected to point d, bulb 1 will __remain lit __go out and bulb 2 will __remain lit __go out.

b. If point d is connected to point e, _____

c. If point e is connected to point f, _____

d. If point f is connected to point a, _____

e. If point a is connected to point c, _____

f. If point a is connected to point d, _____

g. If point a is connected to point e, _____

h. If point b is connected to point d, _____

i. If point b is connected to point e, _____

Step 12: Observe what happens if the additional wire is used to make each of the connections.

a. When point c is connected to point d, bulb 1 __remains lit __goes out and bulb 2 __remains lit __goes out.

b. When point d is connected to point e, _____

c. When point e is connected to point f, _____

d. When point f is connected to point a, _____

e. When point a is connected to point c, _____

f. When point a is connected to point d, _____

g. When point a is connected to point e, _____

h. When point b is connected to point d, _____

i. When point b is connected to point e, _____

Summing Up

1. What do open circuits and short circuits have in common?

_____

_____

2. How are open circuits and short circuits different?

_____

_____

3. Examine the cases in parts B and C when both bulbs went out. Is there anything that *all* those cases have in common?

_____

_____

Electricity: Electric Circuits

Be the Battery

Purpose
In this activity, you will provide energy to an electric circuit using your own muscle power.

Required Equipment and Supplies
Genecon® hand-held generator (or equivalent)
3 miniature bulbs (6-volt flashlight bulbs)
3 miniature bulb sockets
6 connecting wires

Discussion
Batteries last longer in some circuits than they do in others. They last longer when they don't have to "work" so hard. In this activity, *you* will do the work of the battery. That is, you will power a circuit using the hand-held generator. You will learn which circuits are easier to power and which circuits are harder to power. And you'll gain a better appreciation for what batteries and the local power utility do for you all the time!

Procedure
Step 1: Arrange a simple circuit using the generator and a bulb in its socket as shown in figure 1. Gently crank the handle to make the bulb light up. Take care not to crank the generator too quickly and don't give it any sudden jerks or bursts of motion.

Figure 1. The hand-powered circuit

1. When the bulb is lit, how can you make it brighter? Does this require more effort on your part?

_____

Step 2: While cranking the generator and lighting the bulb, have a partner unscrew the bulb from the socket as shown in figure 2.

2. What happens to the cranking effort when the bulb is unscrewed from its socket?

_____

3. When the bulb is removed from the socket, is the resulting circuit an open circuit or a short circuit?

_____

4. Is the electrical resistance in this kind of circuit very high or very low?

_____

Step 3: Remove the generator leads from the bulb terminals and connect them to each other as shown in figure 3.

5. What happens to the cranking effort when the generator leads are connected to each other?

_____

6. When the generator leads are connected directly to each other, is the resulting circuit an open circuit or a short circuit?

_____

Figure 3

7. Is the electrical resistance in this kind of circuit very high or very low?

_____

Step 4: Connect three bulbs in series as shown in figure 4. Gently crank the handle to make the bulb light up. Get a sense of how much effort is needed to power the circuit.

Figure 4. A hand-powered series circuit

Step 5: Connect three bulbs in parallel as shown in figure 5. Gently crank the handle to make the bulb light up. Get a sense of how much effort is needed to power the circuit.

Figure 5. A hand-powered parallel circuit

8. Which circuit is harder to power, the series circuit or the parallel circuit?

_____

Summing Up

1. Which types of circuits are harder to power, those having low electrical resistance or those having high electrical resistance?

2. Which arrangement of three bulbs has more electrical resistance, series or parallel? Justify your answer using observations from the activity.

3. Under which conditions will a battery run down faster, when connected to a high-resistance circuit or when connected to a low-resistance circuit?

4. Which battery would last longer, one connected to three-bulbs series circuit or one connected to three-bulb parallel circuit? (Assume the batteries are identical and the bulbs are identical.)

| CONCEPTUAL PHYSICAL SCIENCE **EXPLORATIONS** | **Activity** |

Electricity: Electric Circuits

Three-Way Switch

Purpose
In this activity, you will explore ways to turn a light bulb on or off from either of two switches.

Required Equipment and Supplies
a 2.5-volt flashlight bulb in a socket
connecting wires
two D-cell batteries with holder, connected in series, or a Genecon hand generator
two single pole double-throw switches

Discussion
Multi-story homes often have one or more stairways with ceiling lights. Two light switches are commonly used to control such a ceiling light. One is placed at the bottom of the stairway; the other is placed at the top. Each switch should be able to turn the light on or off regardless of the setting of the other switch. In this activity you will see how simple but tricky such a common circuit really is!

Procedure
Step 1: Connect a wire from the positive terminal of a 3-volt battery (two 1.5-volt batteries connected positive-to-negative) to the center terminal of a single-pole double-throw switch.

Step 2: Connect a wire from the negative terminal of the same battery to one of the light bulb terminals. Connect the other light bulb terminal to the center terminal of the second switch.

Single pole
double-throw switch

Step 3: Now connect the free terminals of the switches so that the bulb turns on or off from either switch—regardless of the setting of the other switch. Each switch must be in a "down" position: the blade in full electrical contact with one terminal of the switch or the other. The switch shown to the right is not in a "legal" position.

Step 4: When you succeed, draw a simple circuit diagram of your successful circuit.

Summing Up
Will your successful circuit work if you reverse the polarity of the battery? What is your evidence?

_____

_____

CONCEPTUAL PHYSICAL SCIENCE EXPLORATIONS | **Activity**

Magnetism: Magnetic Fields

Magnetic Personality

Purpose
In this activity, you will explore the patterns of magnetic fields around bar magnets in various configurations.

Required Equipment and Supplies
3 bar magnets
iron filings and paper or
a magnetic field projectual (iron filings suspended in oil encased in an acrylic envelope)

Discussion
A magnetic field is a kind of aura that surrounds magnets. Although it can't be seen directly, the overall shape of the field can be seen by the effect it has on iron filings.

Procedure
Step 1: Place a bar magnet on a horizontal surface such as your tabletop. Use the iron filings to see the pattern of the magnetic field.

Iron Filings and Paper Method
Cover the magnet or magnets with a sheet of paper. Then sprinkle iron filings on top of the paper. Jiggle the paper a little bit to help the iron filings find their way into the magnetic field pattern.

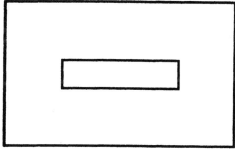

Figure 1

Projectual Method
Mix the iron filings by rotating the projectual. Use the glass rod inside the projectual to help stir the iron filings into a fairly even distribution. Hold the projectual upside down for several seconds before placing it on the magnet or magnets. Take care not to scratch the surface of the projectual by moving it across the magnets once it is in place.

Sketch the field for a single bar magnet in figure 1.

Step 2: Arrange two bar magnets in a line with opposite poles facing each other. Leave about an inch between the poles. Use the iron filings to see the pattern of the magnetic field. Sketch the field for opposite poles in figure 2.

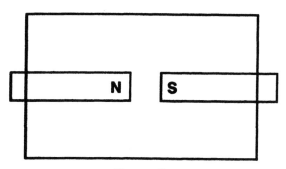

Figure 2

Step 3: Arrange two bar magnets in a line with north poles facing each other. Leave about an inch between the poles. Use the iron filings to reveal the magnetic field. Sketch the field for north poles in figure 3.

Step 4: Predict the pattern for the magnetic field of two south poles facing each other as shown in figure 4. Make a predictive sketch in figure 4.a. After completing your prediction, arrange two bar magnets in a line with south poles facing each other. Leave about an inch between the poles. Use the iron filings to reveal the magnetic field. Sketch the field for south poles in figure 4.b.

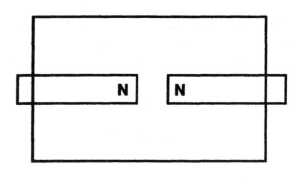

Figure 3

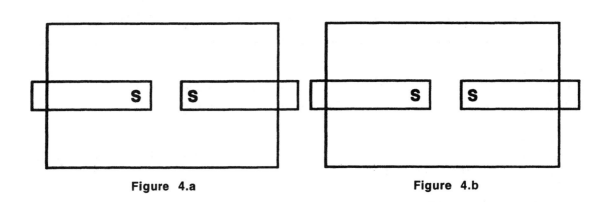

Figure 4.a   **Figure 4.b**

Step 5: Predict the pattern for the magnetic field of two bar magnets parallel to each other as shown in figure 5. Make a sketch on figure 5.a. After completing your prediction, arrange two bar magnets parallel to each other. Use the iron filings to reveal the magnetic field. Sketch the field for two magnets parallel to each other on figure 5.b.

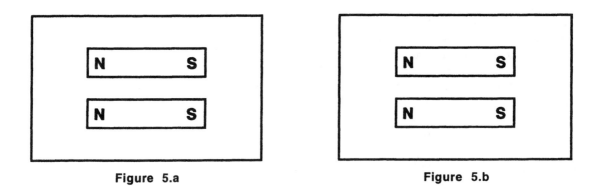

Figure 5.a   **Figure 5.b**

Step 6: Predict the pattern for the magnetic field of two bar magnets anti-parallel to each other as shown in figure 6. Make a sketch on figure 6.a. After completing your prediction, arrange two bar magnets anti-parallel to each other. Use the iron filings to reveal the magnetic field. Sketch the field for two magnets parallel to each other on figure 6.b.

56

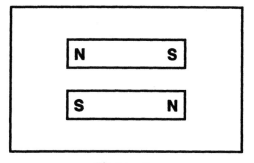

Figure 6.a

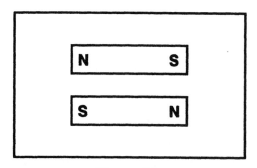

Figure 6.b

Step 7: Can you arrange three bar magnets to create the magnetic field shown in figure 7? If so, how? Is there more than one way to do it?

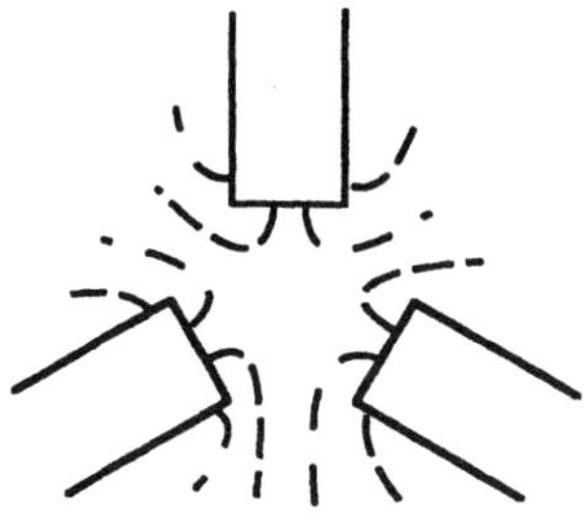

_____

_____

_____

1. Does this pattern show attraction or repulsion?

_____

Figure 7

Step 8: Can you arrange three bar magnets to create the magnetic field shown in figure 8? If so, how? Is there more than one way to do it?

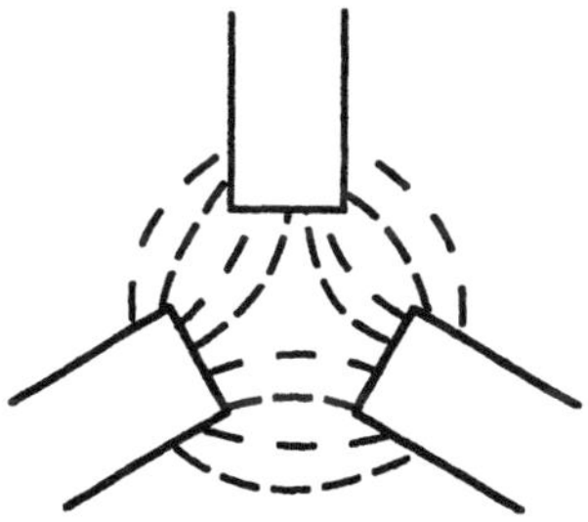

_____

_____

_____

2. Does this pattern show attraction or repulsion?

_____

Figure 8

Summing Up

1. Suppose you see a magnetic field pattern as shown in figure 9. Can you say for sure which pole is north and which pole is south?

_____

_____

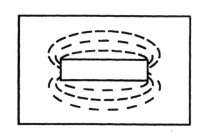

Fig. 9

2. Suppose you see a magnetic field pattern as shown in figure 10. Can you say for sure which pole is north and which pole is south?

_____

_____

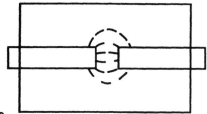

Fig.10

57

3. Suppose you see a magnetic field pattern as shown in figure 11. If pole A is a north pole, what is pole B?

_____

_____

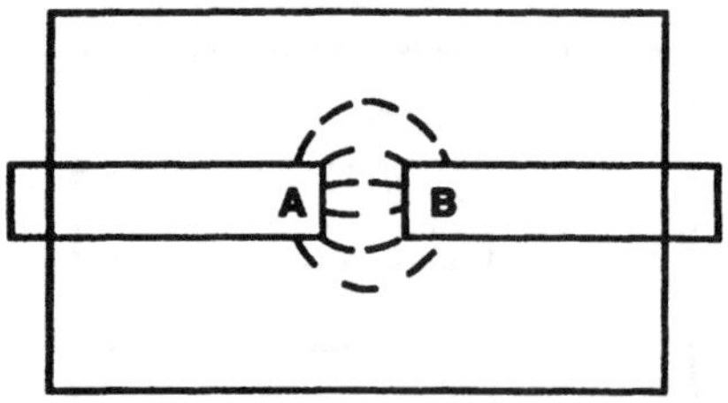

Fig.11

4. Suppose you see a magnetic field pattern as shown in figure 12. If pole C is a north pole, what is pole D?

_____

_____

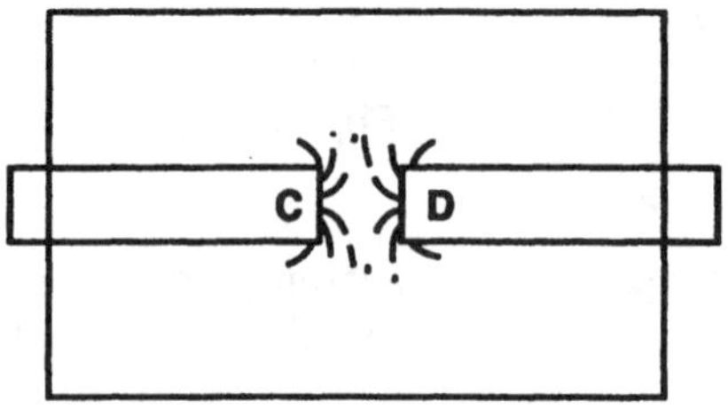

Fig.12

5. Suppose you see a magnetic field pattern as shown in figure 13. If pole E is a north pole, what are poles F, G, H, I, J, K, and L?

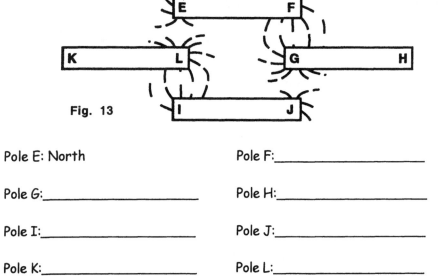

Fig. 13

Pole E: North   Pole F:_____

Pole G:_____   Pole H:_____

Pole I:_____   Pole J:_____

Pole K:_____   Pole L:_____

6. Which of the patterns in figure 14—if either—is/are possible using three bar magnets?

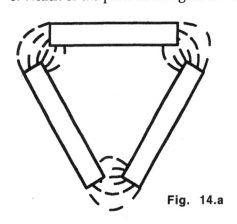

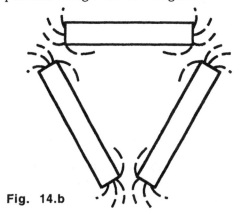

Fig. 14.a   Fig. 14.b

CONCEPTUAL PHYSICAL SCIENCE **EXPLORATIONS**	**Activity**

Waves and Sound: Vibrations

Slow-Motion Wobbler

Purpose
In this activity, you will observe and explore the oscillation of a tuning fork.

Required Equipment and Supplies
variety of tuning forks (low frequency [ex: 40-150 Hz] forks work best for large amplitudes) strobe light, variable frequency

Discussion
The tines of a tuning fork oscillate at a very precise frequency. That's why musicians use them to tune instruments. In this activity, you will investigate their motion with a special illumination system—a *stroboscope*.

Procedure
Strike a tuning fork with a mallet or the heel of your shoe (do *not* strike against the table or other hard object). Does it appear to vibrate? Try it again, this time dipping the tip of the tines just below the surface of water in a beaker. What do you observe?

_____

_____

Now dim the room lights and strike a tuning fork while it is illuminated with a strobe light. For best effect, use the tuning fork with the longest tines available. Adjust the frequency of the strobe so that the tines of the tuning fork appear to be stationary. Then carefully adjust the strobe so that the tines slowly wag back and forth. Describe your observations.

_____

_____

Summing Up
1. What happens to the air next to the tines as they oscillate?

_____

2. Strike a tuning fork and observe how long it vibrates. Repeat, placing the handle against the tabletop or counter. Although the sound is louder, does the *time* the fork vibrates increase or decrease? Explain.

_____

_____

3. What would happen if you struck the tuning fork in outer space?

_____

| CONCEPTUAL PHYSICAL SCIENCE **EXPLORATIONS** | **Activity** |

Waves and Sound: Interference

Sound Off

Purpose
In this demonstration, you will hear a dramatic effect of the interference of sound.

Required Equipment and Supplies
Stereo radio, tape, or CD player with two moveable speakers, one of which has a DPDT (double pole double throw) switch or a means of reversing polarity.

Discussion
Interference is a behavior common to all waves. With water waves we see it in regions of calm where overlapping crests and troughs coincide. We see the effects of interference in the colors of soap bubbles and other thin films where reflection from nearby surfaces puts crests coinciding with troughs. In this activity we'll dramatically experience the effects of interference with sound!

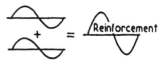

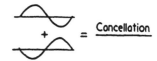

Procedure
Play the stereo player with both speakers in phase (with the plus and minus connections to each speaker the same). Play it in mono mode so the signals of each speaker are identical. Note the fullness of the sound. Now reverse the polarity of one of the speakers (either by physically interchanging the wires or by means of the switch provided). Note the sound is different—it lacks fullness. Some of the waves from one speaker are arriving at your ear out of phase with waves from the other speaker.

Now place the speakers facing each other at arm's length. The long waves are interfering destructively, detracting from the fullness of the sound. Gradually bring the speakers closer to each other. What happens to the volume and fullness of the sound heard? Bring them face to face against each. What happens to the volume now?

Summing Up
1. What happens when to the volume of sound when the face-to-face speakers are switched so both are in phase?

_____

2. Why is the volume so diminished when the out-of-phase speakers are brought together face to face? And why is the remaining sound so "tinny"?

_____

3. What practical applications can you think of for canceling sound?

_____

Reflection and Refraction: Image Formation

Pinhole Image

Purpose
To investigate the operation of a pinhole "lens" and compare it to the eye.

Required Equipment and Supplies
3" x 5" card
straight pin
meterstick

Discussion
The image cast through a pinhole in a pinhole camera has the property of being in clear focus at any distance from the pinhole. That's because the tinyness of the pinhole does not allow overlapping of light rays. (The tinyness also doesn't allow the passage of much light, so pinhole images are normally dim as well.) When a pinhole is placed at the center of the pupil of your eye, the light that passes through the pinhole forms a focused image no matter where the object is located. Pinhole vision, although dim, is remarkably clear. In this activity, you will use a pinhole to see fine detail more clearly.

Procedure
Step 1: Bring this printed page closer and closer to your eye until you cannot clearly focus on it any longer. Even though your pupil is small, your eye does not act like a true pinhole camera because it does not focus well on nearby objects.

Step 2: Poke a single pinhole into a card. Hold the card in front of your eye and read these instructions through the pinhole. Bright light on the print may be required. Bring the page closer and closer to your eye until it is a few centimeters away. You should be able to read the type clearly. Then quickly remove the card and see if you can still read the instructions without the benefit of the pinhole.

Summing Up
Enlist the help of people in your lab who are nearsighted and who are farsighted (if you're not one of them yourself).

1. A farsighted person without corrective lenses cannot see close-up objects clearly. Can a farsighted person without corrective lenses see close-up objects clearly through a pinhole?

_____

2. A nearsighted person without corrective lenses cannot see far-away object clearly. Can a nearsighted person without corrective lenses see far-away objects clearly through a pinhole?

_____

3. Why does a page of print appear dimmer when seen through the pinhole?

_____

CONCEPTUAL PHYSICAL SCIENCE **EXPLORATIONS**	**Activity**

Reflection and Refraction: Image Formation

Pinhole Camera

Purpose
In this activity, you will observe images formed by a simple convex lens and compare cameras with and without a lens.

Required Equipment and Supplies
covered shoebox
25 mm converging lens
glassine paper
aluminum foil
masking tape

Discussion
The first camera used a pinhole opening to let light in. Because the hole is so small, light rays that enter cannot overlap. This is why a clear image is formed on the inner back wall of the camera. Because the opening was small, a long time was required to expose the film sufficiently. A lens allows more light to pass through and still focus the light onto the film. Cameras with lenses require much less time for exposure, and the pictures came to be called "snapshots".

Procedure
Step 1: Construct a camera as shown in Figure A. It is a shoebox with a hole about an inch or so in diameter on one end, some glassine paper taped in the center to act as a screen, and an opening for viewing the screen on the other end. Tape some foil over the lens hole of the box. Poke a pinhole in the middle of the foil. Point the camera toward a brightly illuminated scene, such as the window during the daytime. Light enters the pinhole and falls on the glassine paper. Observe the image of the scene on the glassine paper.

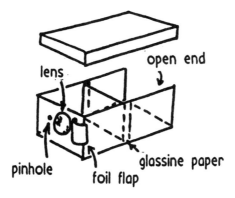

Fig. A

1. Is the image on the screen upside down (inverted)?

_____

2. Is the image on the screen reversed left to right?

_____

Step 2: Now remove the pinhole foil and tape a lens over the hole in the box. You now have a lens camera. Move it around and watch people or other scenery.

3. Is the image on the screen upside down (inverted)?

_____

4. Is the image on the screen reversed left to right?

_____

Step 3: Unlike a lens camera, pinhole cameras focus equally well on objects at practically all distances. Aim the camera lens at nearby objects and see if the lens focuses them.

5. Does the lens focus nearby objects as well as it does on distant ones?

_____

Step 4: Draw a ray diagram as follows. First, draw a ray for light that passes from the top of a distant object through a pinhole and onto a screen. Second, draw another ray for light that passes from the bottom of the object through the pinhole and onto the screen. Then sketch the image created on the screen by the pinhole.

Summing Up

1. Why is the image created by the pinhole dimmer than the one created by the lens?

_____

_____

2. How is a pinhole camera similar to your eye? Do you think that the images formed on the retina of your eye are upside down? Your explanation might include a diagram.

_____

_____

_____

| CONCEPTUAL PHYSICAL SCIENCE **EXPLORATIONS** | **Activity** |

Reflection and Refraction: Reflection

Mirror, Mirror, on the Wall . . .

Purpose
In this activity, you will investigate the minimum size mirror required for you to see a full image of yourself.

Required Equipment and Supplies
large mirror, preferably full length
ruler and masking tape

Discussion

Why do shoe stores and clothier shops have full-length mirrors? Need a mirror be as tall and wide as you for you to see a complete image of yourself?

Procedure
Step 1: Stand about arm's length in front of a vertical full-length mirror. Reach out and place a small piece of masking tape on the image of the top of your head. Now stare at your toes. Place the other piece of tape on the mirror where your toes are seen. Use a meter stick to measure the distance from top of your head to your toes. How does the distance between the pieces of tape on the mirror compare to your height?

_____

Step 2: Now stand about 3 meters from the mirror and repeat. Stare at the top of your head and toes and have an assistant move the tape so that the pieces of tape mark where head and feet are seen. Move further away or closer, and repeat. What do you discover?

_____

Summing Up
1. Does the location of the tape depend on your distance from the mirror?

_____

2. What is the shortest mirror in which you can see your entire image? Do you *believe* it?

_____

Going Further

Try this one if a full-length mirror is not readily available *or* you are a disbeliever! Hold a ruler next to your eye. Measure the height of a common pocket mirror. Hold the mirror in front of you so that the image includes the ruler. How many centimeters of the ruler appear in the image? How does this compare to the height of the mirror?

_____

CONCEPTUAL PHYSICAL SCIENCE **EXPLORATIONS** | **Experiment**

Structure of the Atom: Atomic Size

Thickness of a BB Pancake

Purpose
In this experiment, you will determine the diameter of a BB without actually measuring it. (The abbreviation "BB" stands for "ball bearing.")

Required Equipment and Supplies
75 mL of BB shot
100-mL graduated cylinder
tray
ruler
micrometer

Discussion
This activity distinguishes between *area* and *volume*, and sets the stage for the follow-up experiment *Oleic Acid Pancake*, where you will estimate the size of molecules. To see the difference between area and volume, consider eight wooden blocks arranged to form a single 2 x 2 x 2-inch cube. Since any cube has 6 sides, do you agree the outer surface area will be 6 times the area of one face—24 square inches? The cube form exposes the minimum surface area (which is why buildings in cold areas are most often cubical in shape). Won't the surface area be more in any other configuration? For example, consider an arrangement of a 1 x 2 x 4-inch rectangular block—won't the outer surface area be greater? And if the blocks are spread out to form a stack only one cube thick, 1 x 1 x 8 inches, won't the outer surface area be maximum? Discuss these questions and answers with your lab partners.

The different configurations have different surface areas, but the volume remains constant. The volume of pancake batter is also the same whether it is in the mixing bowl or spread out on a surface (except that on a hot griddle the volume increases because of the expanding gas bubbles that form as the batter cooks). The volume of a pancake equals the surface area of one flat side multiplied by the thickness. If both the volume and the surface area are known, then the thickness can be calculated.

Volume = area x thickness

so simple rearrangement gives: $\text{Thickness} = \dfrac{\text{volume}}{\text{area}}$

Instead of cubical blocks or pancake batter, consider a graduated cylinder that contains BBs. The space taken up by the BBs is easily read as volume on the side of the cylinder. If the BBs are poured into a tray, their volume remains the same. Can you think of a way to estimate the diameter (or thickness) of a single BB without measuring the BB itself? Try it in this activity and see. It will be simply another step smaller to consider the size of molecules.

Procedure

Step 1: Use a graduated cylinder to measure the volume of the BBs. (Note that 1 mL = 1 cm^3.)

Volume = _____ cm^3

Step 2: Carefully spread the BBs out to make a compact layer one pellet thick on the tray. With a ruler, determine the area covered by the BBs. Describe your procedure and show your computations.

Area = _____ cm^2

Step 3: Using the area and volume of the BBs, estimate the thickness (diameter) of a BB. Show your computations.

Estimated thickness = _____ cm

Step 4: Check your estimate by using a micrometer to measure the thickness (diameter) of a BB.

Measured thickness = _____ cm

Summing Up

1. How does your estimate compare to the measurement of the diameter of the BB? Calculate the percentage error (consult the Appendix on how to do this) between the estimated and measured thickness of the BB.

_____

_____

2. Oleic acid is an organic substance that is soluble in alcohol but insoluble in water. When a drop of oleic acid is placed in water, it usually spreads out over the water surface to create a *monolayer*, a layer that is one molecule thick. From your experience with BBs, describe a method for estimating the size of an oleic acid molecule.

_____

_____

_____

Name_____ Period_____ Date_____

Structure of the Atom: Atomic Size

Oleic Acid Pancake

Purpose
In this experiment, you will estimate the size of a single molecule of oleic acid.

Required Equipment and Supplies
tray
water
chalk dust or lycopodium powder
eyedropper
oleic acid solution (5 mL oleic acid in 995 mL of ethanol)
10-mL graduated cylinder

Discussion
During this experiment you will estimate the *diameter* of a single molecule of oleic acid! The procedure for measuring the diameter of a molecule will be much the same as that of measuring the diameter of a BB in the previous activity. The diameter is calculated by dividing the volume of the drop of oleic acid by the area of the *monolayer* film that is formed. The diameter of the molecule is the depth of the monolayer.

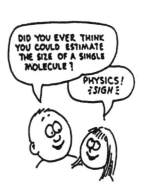

Volume = area x depth

$$\text{Depth} = \frac{\text{volume}}{\text{area}}$$

Procedure
Step 1: Pour water into the tray to a depth of about 1 cm. So that the acid film will show itself, spread chalk dust or lycopodium power very lightly over the surface of the water; too much will hem in the oleic acid.

Step 2: Using the eyedropper, gently add a single drop of the oleic acid solution to the surface of the water. When the drop touches the water, the alcohol in it will dissolve in the water, but the oleic acid will not. The oleic acid spreads out to form a nearly circular patch on the water. Measure the diameter of the oleic acid patch in several places, and compute the average diameter of the circular patch.

Average diameter =_____ cm

The average radius is, of course, half the average diameter. Now compute the area of the circle ($A = \pi r^2$).

Area of circle = _____ cm²

Step 3: Count the number of drops of solution needed to occupy 1 mL (or 1 cm³) in the graduated cylinder. Do this three times, and find the average number of drops in 1 cm³ of solution.

Number of drops in 1 cm³ = _____

Divide 1 cm^3 by the number of drops in 1 cm^3 to determine the volume of a single drop.

Volume of single drop = _____ cm^3

Step 4: The volume of the oleic acid alone in the circular film is much less than the volume of a single drop of the solution. The concentration of oleic acid in the solution is 5 mL per liter of solution. Every cubic centimeter of the solution thus contains only $\frac{5}{1000}$ cm^3, or 0.005 cm^3, of oleic acid. The volume of oleic acid in one drop is thus 0.005 of the volume of one drop. Multiply the volume of a drop by 0.005 to find the volume of oleic acid in the drop. This is the volume of the layer of acid in the tray.

Volume of oleic acid = _____ cm^3

Step 5: Estimate the diameter of an oleic acid molecule by dividing the volume of oleic acid by the area of the circle.

Diameter = _____ cm

The diameter of an oleic acid molecule as obtained by this method is good, but not precise. This is because an oleic acid molecule is not spherical, but rather elongated like a hot dog. One end is attracted to water, and the other end points away from the water surface. The molecules stand up like people in a puddle! So the estimated diameter is actually the estimated length of the short side of an oleic acid molecule.

Summing Up
1. What is meant by a *monolayer*?

_____

2. Why is it necessary to dilute the oleic acid for this experiment? Why alcohol?

_____

3. The shape of oleic acid molecules is more like that of a hot dog than a sphere. Furthermore, one end is attracted to water (*hydrophilic*) so that the molecule stands up on the surface of water. Assume an oleic molecule is 10 times longer than it is wide. Then estimate the volume of one oleic acid molecule.

_____

_____

| CONCEPTUAL PHYSICAL SCIENCE **EXPLORATIONS** | **Activity** |

Radioactivity: Radioactive Half-life

Get a Half-life!

Purpose
In this activity, you will simulate radioactive decay half-life.

Required Equipment and Supplies
25 small color-marked cubes per group (one side red, two sides blue, three sides blank). Spray-painted sugar cubes work well. Multifaceted dice may also be used.

Discussion
The rate of decay for a radioactive isotope is measured in terms of **half-life**—the time for one half of a radioactive quantity to decay. Each radioactive isotope has its own characteristic half-life (Table 1). For example, the naturally occurring isotope of uranium, uranium-238, decays into thorium-234 with a half-life of 4,510,000,000 years. This means that only half of an original amount of uranium-238 remains after this time. After another 4,510,000,000 years, half of this decays leaving only one-fourth of the original amount remaining. Compare this with the decay of polonium-214, which has a half-life of 0.00016 *seconds*. With such a short half-life, any sample of polonium-214 will quickly disintegrate.

Table 1

Isotope	Half-life
Uranium-238	4,510,000,000 years
Plutonium-239	24,400 years
Carbon-14	5,730 years
Lead-210	20.4 years
Bismuth-210	5.0 days
Polonium-214	0.00016 seconds

The half-life of an isotope can be calculated by the amount of radiation coming from a known quantity. In general, the shorter the half-life of a substance, the faster it decays, and the more radioactivity per amount is detected.

In this activity, you will investigate three hypothetical substances, each represented by a color on the face of a cube. The first substance, represented by a given color, is marked on only one side of the cube. The second substance, represented by a second color, is marked on two sides of the cube, and the third substance, represented by a third color (or lack thereof), is marked on the remaining three sides. Rolling a large number of these identically painted cubes simulates the process of decay for these substances. As a substance's color turns face up, it is considered to have decayed and is removed from the pile. This process is repeated until all of the cubes have been removed. Since the color of the first substance is only on one side, this substance will decay the slowest (because its color will fall face up least frequently and it will stay in the game longer). The second substance, marked on two sides, will decay faster requiring fewer rolls before all the cubes are removed. The third substance, marked on three sides, will decay the fastest. After tabulating and graphing the numbers of cubes that decay in each roll for these simulated substances, you will be able to determine their half-lives.

73

Procedure

Step 1: Shake the cubes in a container and roll them onto a flat surface.

Step 2: Count the one-side color faces that are up and record this number under "Removed" in the Data Table.

Step 3: Remove the one-side color cubes in a pile off to the side.

Step 4: Gather the remaining cubes back into the container and roll them again.

Step 5: Repeat steps 2-4 until all cubes have been counted, tabulated and set aside.

Step 6: Repeat steps 1-5 removing cubes that show the two-side color faces up.

Step 7: Repeat steps 1-5 removing cubes that show the three-side color faces up.

Data Table

Throw	First Substance (One-side color)		Second Substance (Two-side color)		Third Substance (Three-side color)	
	Removed	Remaining	Removed	Remaining	Removed	Remaining
Initial Count						
1						
2						
3						
4						
5						
6						
7						
8						
9						
10						
11						
12						
13						
14						
15						
16						
17						
18						
19						
20						
21						
22						
23						
24						
25						

Step 8: Plot the number of cubes remaining versus the number of throws for each substance on the following graph. Use a different color or line pattern to graph the results for each substance. For each substance, draw a single smooth line or curve that approximately connects all points. **Do not connect the dots!** Indicate your color or line pattern code below the graph.

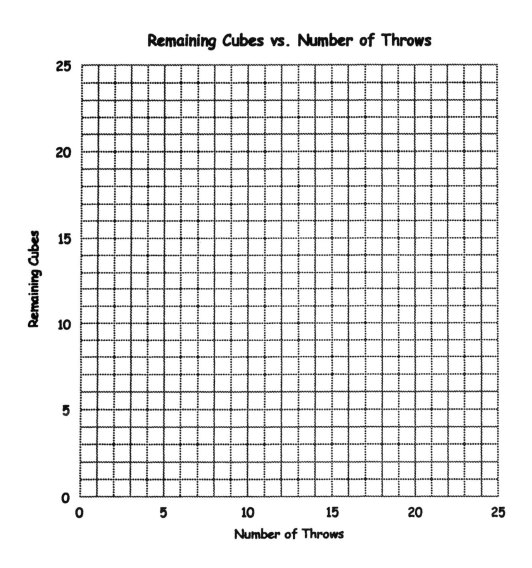

First substance (one side on cube) color or line pattern:

Second substance (two sides on cube) color or line pattern:

Third substance (three sides on cube) color or line pattern:

Summing Up

1. How many rolls did it take for the number of each colored cube to be reduced by half? These are your half-life readings.

   One-side color _____   Two-side color _____   Three-side color _____

2. The half-life of a decaying substance is measured in units of time. What is the unit of half-life used in this simulation?

   _____

3. In each case, how many rolls did it take to remove all of the cubes?

   One-side color _____   Two-side color _____   Three-side color _____

4. Which of these hypothetical substances would be the most harmful?

   _____

5. How might you simulate the radioactive decay of a substance that decays into a second substance that also decays?

   _____

   _____

6. Is it possible to estimate the half-life of a substance in a single throw? How accurate might this estimate be?

   _____

   _____

7. Are your lines in the graph for step 8 fairly straight or do they curve? Do these lines correspond to a constant or non-constant rate of decay?

   _____

   _____

8. a. Substance X has a half-life of 10 years. If you start with 1000 g, how much will be left after:

      i. 10 years? _____

      ii. 20 years? _____

      iii. 50 years? _____

      iv. 100 years? _____

   b. Will this sample of substance X ever totally disappear? If so, estimate how soon. If not, explain.

   _____

CONCEPTUAL PHYSICAL SCIENCE **EXPLORATIONS** | **Activity**

Nuclear Fission and Fusion: Chain Reaction

Chain Reaction

Purpose
In this activity, you will simulate a simple chain reaction

Required Equipment and Supplies
100 dominoes
large table or floor space
stopwatch

Discussion
Give your cold to two people who in turn give it to two others who in turn do the same on down the line and before you know it, everyone in class is sneezing. You have set off a chain reaction. Similarly, when one electron in a photomultiplier tube in certain electronic instruments hits a target that releases two electrons that in turn do the same on down the line, a tiny input produces a large output. When one neutron triggers the release of two or more neutrons in a piece of uranium, and the triggered neutrons trigger others in succession, the results can be devastating. In this activity, we'll explore this idea.

Procedure
Step 1: Set up a strand of dominoes about half a domino length apart in a straight line. Gently push the first domino over, and measure how long it takes for the entire strand to fall over (just like those television commercials).

Step 2: Arrange the dominoes as in Figure 1, so that when one domino falls, another one or two are toppled over. These topple others in chain reaction fashion. Set up until you run out of dominoes or table space. When you finish, push the first domino over and watch the reaction. Notice the number of the falling dominoes per second at the beginning versus the end.

Fig. 1

Summing Up
1. Which reaction, wide-spaced or close-spaced dominoes, took a shorter time?

_____

_____

2. How did the number of dominoes being knocked over per second change for each reaction?

_____

_____

3. What caused each reaction to stop?

_____

_____

4. Now imagine that the dominoes are the neutrons released by uranium atoms when they fission (split apart). Neutrons from the nucleus of a fissioning uranium atom hit other uranium atoms and cause them to fission. This reaction continues to grow if there are no controls. Such an uncontrolled reaction occurs in a split second and is called a *nuclear explosion*. How is the domino chain reaction similar to the nuclear fission process?

_____

_____

_____

_____

5. How is the domino reaction dissimilar to the nuclear fission process?

_____

_____

_____

_____

Name_____ Period_____ Date_____

CONCEPTUAL PHYSICAL SCIENCE EXPLORATIONS | **Experiment**

Elements of Chemistry: Qualitative Analysis

Mystery Powders

Purpose

In this experiment, you will identify nine common household chemicals by using qualitative analysis, a method of determining a substances identity by subjecting it to a series of tests.

Required Equipment and Supplies

Household chemicals and formulae
  cornstarch, $(C_6H_{10}O_5)_n$
  white chalk, $CaCO_3$
  plaster of Paris, $2CaSO_4 \cdot H_2O$
  washing soda, Na_2CO_3
  epsom salt, $MgSO_4 \cdot 7H_2O$
  baking soda, $NaHCO_3$
  boric acid, H_3BO_3
  table sugar, $C_{12}H_{22}O_{11}$
  table salt, NaCl

Test reagents
  tincture of iodine
  phenolphthalein
  white vinegar
  rubbing (isopropyl) alcohol (70%)
  sodium hydroxide (0.3 M)

Equipment
  test tubes*
  eye dropper
  safety goggles
  beaker
  spatula

Note to teacher: glass or plastic well plates may be substituted for test tubes.

Safety!

You will be working with unknown chemicals. Handle them carefully and never, ever taste them. Some may taste sweet but others may burn your tongue! If you are unsure about any procedure, ask your teacher. Avoid spillage. Use small quantities—no more than is required for each test. Never place unused unknown back in the reagent bottle, because this can contaminate the stock material. Consult your instructor for proper disposal.

Discussion

You are given nine vials and each contains a white powder, which is a common household chemical. Your task is to identify these unknowns based upon their different physical and chemical properties. For this experiment you should develop a qualitative analysis scheme, such as shown in Figure 1, to show how the chemicals can be systematically identified.

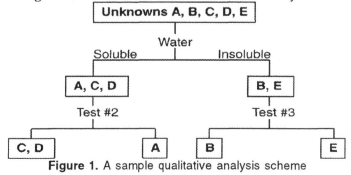

Figure 1. A sample qualitative analysis scheme

Procedure

Using the tests outlined below prepare a qualitative analysis scheme that should permit you to unequivocally determine the identity of each unknown. This scheme should be prepared before conducting this experiment. It is recommended that you start with Test #1 (Solubility in Water). You may then choose your own order of testing. Some orders are more efficient than others. Try to develop a scheme that minimizes the number of tests that must be conducted. Use Table 1 as a guide to the physical and chemical properties of the household chemicals to be identified.

Tests

1. Solubility in Water

Place a pea-sized portion of the unknown solid in a test tube and add about 5 mL of water. To mix the contents, hold the top of the test tube securely between your thumb and index finger and gently slap the bottom of the test tube with the index finger of your opposite hand. Be careful so that none of the solution spills out. Consult Table 1 for behaviors of each of the household chemicals.

2. Tincture of Iodine

This test may be used for any unknown that is insoluble in water. Add a few drops of tincture of iodine to the unknown as it sits undissolved in water. A deep blue color forms as the iodine complexes with cornstarch. A brownish color will appear for all other unknowns insoluble in water.

3. Phenolphthalein

This test may be used for any unknown that is soluble in water. Add a drop of phenolphthalein to the dissolved unknown and a bright pink color will result if the solution is very alkaline. This test is positive for washing soda, Na_2CO_3, which turns a light pink.

4. White Vinegar

This test is applicable to all unknowns. Add a few drops of vinegar to the unknown, in either a dissolved or undissolved state. The formation of bubbles is a sign of the carbonate ion [CO_3^{2-}], which decomposes to gaseous carbon dioxide upon treatment with an acid (vinegar). This test is positive for white chalk, $CaCO_3$, washing soda, Na_2CO_3, and baking soda, $NaHCO_3$.

5. Sodium Hydroxide (0.3 M):

This is a specific test for magnesium sulfate, $MgSO_4$, which is found in epsom salts. If the unknown dissolves in water and forms an insoluble precipitate when treated with this test reagent, then the unknown contains magnesium sulfate.

6. Hot Water

All the water soluble unknowns become markedly more soluble in warmer water with the exception of sodium chloride, $NaCl$ (table salt). This test, therefore, is specific to sodium chloride. Place several pea-sized portions of the unknown solid in a test tube along with 5 mL of water. Heat in a hot water bath held at about 60°C. No marked improvement in solubility suggests that the unknown may be sodium chloride.

7. Rubbing Alcohol:

This test may be used for any unknown that is soluble in water. Place a pea-sized portion of the unknown solid in a test tube and add about 7 mL of rubbing alcohol (isopropyl, 70%). Only two substances should dissolve readily: epsom salt, $MgSO_4$, and boric acid, H_3BO_3. Four substances will not dissolve readily: baking soda, $NaHCO_3$, washing soda, Na_2CO_3, sugar, $C_{12}H_{22}O_{11}$, and salt, $NaCl$.

Table 1. Properties of nine household chemicals

Household Chemical	Reactions to Tests						
	#1 Solubility	#2 Iodine	#3 pH >7	#4 Vinegar	#5 Hydroxide	#6 Hot water	#7 Alcohol
cornstarch	insoluble	positive	—	negative	—	—	—
white chalk	insoluble	negative	—	positive	—	—	—
plaster of Paris	insoluble	negative	—	negative	—	—	—
washing soda	soluble	—	positive	positive	negative	greater sol.	insoluble
epsom salt	soluble	—	negative	negative	positive	greater sol.	soluble
baking soda	soluble	—	negative	positive	negative	greater sol.	insoluble
boric acid	soluble	—	negative	negative	negative	greater sol.	soluble
table sugar	soluble	—	negative	negative	negative	greater sol.	insoluble
table salt	soluble	—	negative	negative	negative	no effect	insoluble

Summing Up

1. Identify your unknowns.

   Corn starch was unknown: _____   White chalk was unknown: _____

   Plaster of Paris was unknown:_____   Washing soda was unknown:_____

   Epsom salt was unknown:_____   Baking soda was unknown:_____

   Boric acid was unknown:_____   Table sugar was unknown:_____

   Table sugar was unknown:_____

2. Which of the tests used in this experiment measure physical properties and which measure chemical properties?

   _____

3. Sugar is a chemical, but it is also a food. Is this contradictory? Briefly explain.

   _____

4. How many individual tests did you have to perform to identify all of the unknowns? (Review your qualitative analysis scheme or simply count your number of dirty test tubes.)

   _____

   _____

81

Name_____ Period_____ Date_____

| CONCEPTUAL PHYSICAL SCIENCE **EXPLORATIONS** | **Experiment** |

Mixtures: Separation of a Mixture

Salt and Sand

Purpose
In this experiment, you will develop a laboratory procedure for separating components in a mixture of salt and sand, carry out the separation, and calculate the mass percent of salt and of sand in the mixture.

Required Equipment and Supplies
various salt-sand samples (about 10 grams each)
safety goggles
various pieces of laboratory equipment,
depending on the procedure you develop

Discussion
Components of a mixture can be separated by their different physical properties. In this experiment you are to isolate salt and sand from a mixture. Knowing the mass of the isolated salt and the mass of the mixture, you can calculate the percent composition of the salt that was in the mixture. Likewise, knowing the mass of the isolated sand and the mass of the mixture, you can calculate the percent composition of the sand that was in the mixture. Your teacher will show you how to correctly use and care for laboratory equipment you may use, such as balances and hot plates. The procedure you follow for finding the percent compositions, however, is to be created by you. Before you do anything, therefore, sit down and write out what you think might be a good procedure to follow. For ideas, you might wish to discuss the possibilities with your classmates as well as your teacher. Label this procedure as your "Proposed Procedure".

Before you begin to work in the laboratory, show your written step-by-step proposed procedure to you teacher for final approval. Don't be surprised if you have to modify your procedure as you move along. This is typical. You should also create, therefore, an "Actual Procedure" that shows what was actually done.

Safety!
Significant changes to your proposed procedure must be approved by your teacher. If you have approval to be working with a flame, remove all combustible materials, such as paper towels, from your work station. Eye protection is advised because it is not just what you're doing that poses a hazard. For example, others around you might accidentally shatter glassware, which could send glass pieces towards your face.

Procedure
Step 1: On a separate piece of paper or in a journal, write out a step-by-step procedure that you think will best allow you to find the amount of salt and sand within a mixture. Have your teacher approve of this procedure before moving on. Safety concerns and necessary equipment should be checked. In developing your procedure, keep in mind that you have to end up with the salt and the sand in separate containers so that you can measure the mass of each component.

Step 2: Obtain a sample of salt and sand from your teacher and write down its reference number by your procedures and within question 1 of Summing Up.

Step 3: Begin following your proposed procedure, while recording your actual procedure. Write down numerical measurements, such as mass, neatly within one or more data tables. Include units with any number you write down.

Step 4: Use the following equations to calculate the mass percents of salt and sand in your mixture:

$$\text{mass percent of salt} \ = \ \frac{\text{mass of salt}}{\text{mass of entire sample}} \ \times \ 100\%$$

$$\text{mass percent of sand} \ = \ \frac{\text{mass of sand}}{\text{mass of entire sample}} \ \times \ 100\%$$

Because the mixture contains only two components, the total of the two percentages should be 100%. In reality, various errors may occur such that these two percentages *do not* add up to 100%.

Step 5: Obtain the true values for the mass percents from your teacher.

Summing Up

1. Enter your sample's identification number in this box:

2. What were the mass percents of salt and sand in your sample that you found through your procedure?

mass percent
of salt:

mass percent
of sand:

3. What were the true mass percents of salt and sand as reported to you by your teacher after you had completed your procedure?

mass percent
of salt:

mass percent
of sand:

4. What do you think was your most significant source of error in determining the mass percent of salt?

_____

_____

5. What do you think was your most significant source of error in determining the mass percent of sand?

_____

_____

CONCEPTUAL PHYSICAL SCIENCE **EXPLORATIONS**	**Experiment**

Mixtures: Sugar Content of Soft Drinks

Sugar Soft

Purpose

In this experiment, you will determine the sugar content of commercially sold soft drinks using a home-built hydrometer consisting of a 9-inch plastic pipet and nut.

Required Equipment and Supplies

9-inch plastic pipet
1/2-inch nut
sugar solutions (4%, 8%, 12%, 16%)
50-mL graduated cylinder
ruler (cm)
graph paper
various soft drinks (Try fruit juices and diet drinks as well!)

Discussion

A hydrometer is a flotation device used to measure the density of a liquid. The greater the density of the liquid, the higher the hydrometer floats. In this exploration, how high the hydrometer floats in four standard sugar solutions will be measured. The greater the sugar content of the solution, the greater its density, hence, the higher the hydrometer floats. A *calibration curve* will be graphed showing the height of the hydrometer on the y-axis and the concentration of sugar on the x-axis. How high the hydrometer floats in various commercially prepared soft drinks, which are essentially sugar solutions with small amounts of other materials, will then be measured. Using the calibration curve, the sugar content of each soft drink may be estimated.

How high the hydrometer floats out of the water is the distance between its tip and the liquid surface (use units of millimeters). Make sure that the hydrometer is not held to the sides of the container—it should float as vertically as possible. The hydrometer must be rinsed and dried before each testing. Also, carbonated beverages must be "decarbonated" because bubbles will collect on the hydrometer affecting its buoyancy. Your teacher will have decarbonated beverages by boiling them and then allowing them to cool.

Procedure

Part A: Construction and Calibration of a Simple Hydrometer

Step 1: Fill the 9-inch pipet about one-half full with water and invert. All the water should run into the bulb.

Step 2: Slip the nut onto the stem end and allow it to rest on the "shoulders" of the bulb as shown to the right.

Step 3: Test your hydrometer by placing it, bulb end down, in the 50-mL graduated cylinder containing 50 mL of water. The pipet should float with about 2.5 cm of the stem sticking out above the water. If it sticks out much more or less than this, either add water to the pipet or remove water from it.

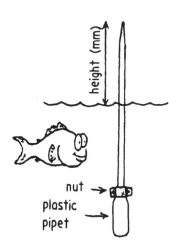

A simple hydrometer

Step 4: When you have adjusted the amount of water in the pipet bulb so that the stem sticks out about 2.5 cm, measure the height of the stem above the surface of the water in the graduated cylinder. Measure to the nearest millimeter, and record the height on Data Table 1.

Step 5: Remove the hydrometer from the graduated cylinder. *Being careful not to let any water spill out of the hydrometer or any water get in*, rinse the outside surface of the hydrometer well and dry completely.

Step 6: Empty the graduated cylinder, rinse well, and dry completely.

Step 7: Add 50 mL of the 4% sugar solution to the graduated cylinder, place the hydrometer in the solution—bulb end down—and measure and record the height of the stem above the surface of the solution.

Step 8: Repeat steps 5, 6, and 7 for 8%, 12%, and 16% sugar solutions and enter all your data into Data Table 1.

Step 9: Plot the information in Data Table 1 into the large graph presented at the end of this exploration. Draw a straight line that best represents all the data points, as the example in Figure 1 shows.

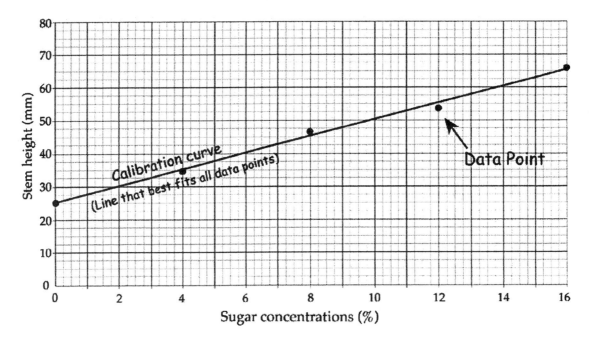

Figure 1. The calibration curve shown here is a line that best represents all of the data points. Note that the calibration curve does not necessarily touch all points.

Part B: Determination of Sugar Content in Beverages

Step 1. Obtain soft drink and/or juice samples from your teacher.

Step 2. Following the procedure given in Part A, measure the stem height for each of your beverages and record this information in Data Table 2.

Step 3. Use the calibration curve you prepared in Part A for your particular hydrometer to determine the sugar concentration in each of your beverages. To do this, find the point on the vertical axis that corresponds to the hydrometer stem height for your beverage. Figure 2 shows an example where the stem height is 52.5 mm. Draw a horizontal line that runs rightward from this point until it intersects the calibration curve. Then, drop a vertical line from the intersection point down to the horizontal axis. The value at the point where this line intersects the horizontal axis—about 10.75% in the example shown here—is the sugar concentration of your beverage.

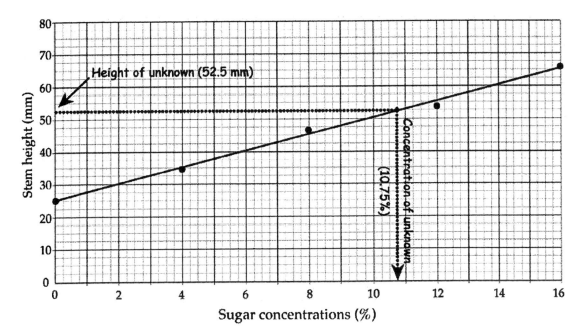

Figure 2. According to this example calibration curve, a stem height of 52.5 mm translates into a sugar concentration of about 10.75%.

Step 4. Record the concentrations of all your beverages in Data Table 2.

Data Table 1

Concentration of Sugar Solution (%)	Height of Hydrometer Above Liquid Surface (mm)
0 (plain water)	
4	
8	
12	
16	

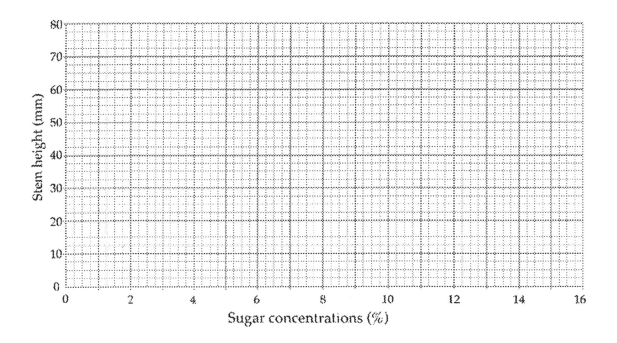

Data Table 2

Brand Name or Type of Beverage	Height of Hydrometer Above Liquid Surface (mm)	Sugar Concentration of Beverage (%)

Chemical Bonding: Collection of a Gas

Foamy Bubble Round-Up

Purpose
In this activity, you will isolate gaseous carbon dioxide by water displacement.

Required Equipment and Supplies
baking soda
vinegar
2 250-mL Erlenmeyer flasks
1000-mL beaker
2 rubber stoppers (one with glass rod inserted through it)
tubing with a paper clip inserted in one end
ring stand with clamp
safety goggles

Discussion
A common method of collecting a gas produced from a chemical reaction is by the displacement of water. Bubbles of the gas are directed into an inverted flask filled with water. As the gas rises into the container, it displaces water. Once all the water is displaced, the flask may be sealed with a stopper. The physical and chemical properties of the gas can then be investigated. In this activity, baking soda (sodium bicarbonate, $NaHCO_3$), and vinegar (5% acetic acid, CH_3COOH), will be reacted to form gaseous carbon dioxide, CO_2, which will be collected by water displacement. The other products of this reaction, water, H_2O, and sodium acetate, $CH_3COO^-Na^+$, form a liquid phase that remains in the reaction vessel.

$NaHCO_3$	+	CH_3CO_2H	$\Rightarrow$	CO_2	+	H_2O	+	CH_3CO_2Na
sodium bicarbonate		acetic acid		carbon dioxide		water		sodium acetate

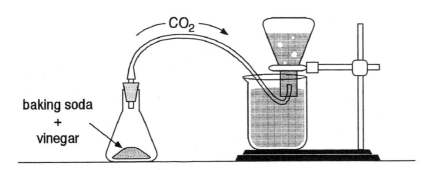

Figure 1. Baking soda (sodium bicarbonate) and vinegar (acetic acid) react to form gaseous carbon dioxide, CO_2.

Procedure
Step 1: Assemble the set-up as shown in Figure 1 or one presented to you by your teacher. A 250-mL Erlenmeyer flask is equipped with a rubber stopper through which a glass tube has been inserted. The glass tube is attached to a plastic tube that is forced into a J-shape on the opposite end by the bending of an inserted paper clip. A second 250-mL Erlenmeyer flask is

filled with water and inverted into a 1000-mL beaker filled with 700 mL of water. The J-shaped end of the plastic tubing is then fixed below the lip of the inverted flask.

Step 2: Remove the stopper from the upright Erlenmeyer flask and add about a tablespoon of baking soda (sodium bicarbonate) followed by a capful of vinegar (5% acetic acid). Allow the vigorous bubbling to settle down. Add a second capful of vinegar and immediately stopper the upright flask. This will cause bubbles to fill the inverted flask. When bubbling stops, add an additional capful of vinegar to collect additional bubbles (carbon dioxide). Continue in this manner until the inverted beaker has been filled with carbon dioxide. *(Caution: Add vinegar only by the capful. Froth may overflow the flask if you add too much vinegar all at once.)*

Step 3: Stopper the inverted flask tightly to seal the carbon dioxide. Remove the flask from the clamp and turn it right side up with the stopper still in place.

Step 4: Light a wooden match and hold it into the carbon dioxide-containing flask. The flame should quickly extinguish. You might also light a small candle and slowly pour the carbon dioxide out of the flask onto the candle to extinguish the candle. This works because the carbon dioxide is heavier than air and will thus pour out of the flask.

Going Further
Step 5: Weigh the mass of an empty and dry 2-liter bottle with its cap. The mass of the air inside this bottle is about 2.3 grams. Subtract the mass of the air inside the bottle from the mass you measure to get the mass of only the bottle and its cap.

Step 6: Insert the J-hook end of the tube into the upright bottle and pipe lots of carbon dioxide into the bottle. Make sure that the tube is dry and that no water drops get into the bottle. Because the CO_2 is heavier than air, it will settle to the bottom and displace the air, which will escape out the top. Continue filling the bottle with CO_2 until a match is extinguished when placed into the opening.

Step 7: Cap the bottle and measure its mass. Subtract the mass of the bottle and its cap from the mass you measure to get the mass of the carbon dioxide it contains. Calculate the density of the carbon dioxide by dividing its mass by volume. Compare your experimental value to the actual density of carbon dioxide, which is 1.80 grams per liter.

Your calculated density of carbon dioxide:	The actual density of carbon dioxide:
_____	1.80 grams per liter

Summing Up
1. How certain are you that an empty 2-liter plastic soda bottle contains a volume of exactly 2 liters? How would you find out for sure?

_____

_____

2. Why was the flask containing baking soda not stoppered after the first capful of vinegar was added?

_____

_____

3. Why was the large beaker not initially filled to the brim?

_____

Name_____ Period_____ Date_____

Chemical Bonding: Molecular Models

Molecules by Acme

Purpose
In this experiment, you will build models of various molecules. First, you will be given a chemical formula. Then, based upon some rules for how atoms bond, you will piece together a molecular model.

Required Equipment and Supplies
molecular modeling kits. Students should be able to work in groups of 2 or 3.

Discussion
The 3-dimensional shapes of molecules can be envisioned with the use of molecular models. There are certain things you must know, however, before you can assemble an accurate model of a molecule. First, you need to know what types of atoms make up the molecule and also their relative numbers. This information is given by the molecule's chemical formula. The chemical formula for water, H_2O, for example, tells us that each and every water molecule is made of two hydrogen atoms and one oxygen atom. The second thing you need to know is how the atoms of the molecule fit together. For a water molecule there are several possibilities. One hydrogen atom might be bonded to both the second hydrogen atom and the oxygen atom (Figure 1a), or all three atoms might be bonded in the shape of a three-member ring (Figure 1b). We find through chemistry, however, that there are specific ways in which different types of atoms bond. Hydrogen, for example, forms only one bond, while oxygen forms two bonds. Knowing this we can build a more reasonable model of a water molecule where both hydrogen atoms are bonded once to a central oxygen atom (Figure 1c).

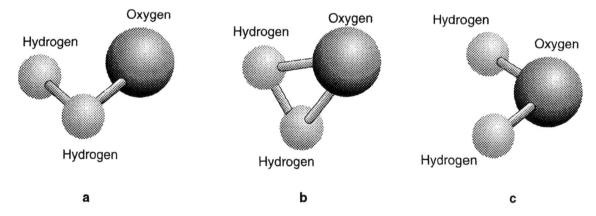

a   b   c

Figure 1. Which version of H_2O is correct?

The shape of a molecule is largely responsible for the physical and chemical properties of the molecule. If water were linear like carbon dioxide, for example, its boiling point would be close to that of carbon dioxide, -78°C, which would mean that the Earth's oceans would be gaseous.

The Molecular Model

Molecular modeling kits represent different types of atoms with different colored pieces. Hydrogen atoms, for example, are typically represented by white pieces and oxygen atoms by red pieces. Also, the number of times an atom prefers to bond is indicated by the number of times the piece is able to attach to other pieces. Hydrogen atoms, for example, bond only once, so hydrogen pieces have only one site for attachment. Similarly, oxygen pieces have two sites for attachment. To build a molecule, the pieces representing atoms are connected by sticks (or springs). You will know when you have built a correct structure for a molecule when each atom is bonded the appropriate number of times (Table 1). In some instances, you may find it necessary to form multiple bonds between atoms. Consult the instructions to your modeling kit or your instructor to see how this is done. This is a hands-on, play as you investigate laboratory. Enjoy!

Table 1

Type of Atom	Atomic Symbol	Typical Color cf Sphere	Number of Bonds* [holes]
Hydrogen	H	White	1
Carbon	C	Black	4
Nitrogen	N	Blue	3
Oxygen	O	Red	2
Chlorine	Cl	Green	1

* The number of times an atom tends to bond is related to its position in the periodic table. Consider, for example, the relative positions of carbon, nitrogen, oxygen and chlorine. That these atoms are in adjacent columns and prefer 4, 3, 2 and 1 bonds, respectively, is not a coincidence. The periodic table is more than a table of facts. With further study you will find that the periodic table is highly organized - and a lot like a roadmap, for much information about an element can be told merely from its position.

Procedure

Step 1: Get a set of models. You may wish to work with a partner. The sets may not contain equal numbers of pieces, so you may occasionally need to borrow from other groups.

Step 2: Determine which colors should be used to represent the following elements: hydrogen, carbon, nitrogen, oxygen, chlorine, and iron. The number of bonds they are able to form should be as listed in Table 1. Enter the actual colors of the pieces you use in Table 2.

Step 3: Build models of each of the following molecules. Avoid forming triangular rings made of 3 atoms. They are strained and less stable. Check your structure with your teacher—in several cases there is more than one possibility. Complete Table 3 by drawing an accurate representation of your structure (follow the example of the water molecule.) Answer the end-of-activity questions using these models.

1. Hydrogen gas.................H_2
2. Oxygen gas ....O_2
3. Nitrogen gas...................N_2
4. Water ....H_2O
5. Hydrogen Peroxide ....H_2O_2
6. Ammonia.......................NH_3
7. Methane........................CH_4
8. Dichloromethane ....CH_2Cl_2

9. Chloroethanol ....C_2H_5ClO
10. Carbon Dioxide.............CO_2
11. Acetylene.....................C_2H_2
12. Ethanol.......................C_2H_6O
13. Acetic Acid...................$C_2H_4O_2$
14. Benzene.......................C_6H_6
15. Iron (III) Oxide.............Fe_2O_3

Table 2.

Element:	Hydrogen [H]	Carbon [C]	Nitrogen [N]	Oxygen [O]	Chlorine [Cl]	Iron (III)* [Fe]
Color:						

* For iron, choose a piece that is able to bond in 6 directions at angles of 90 degrees.
The iron will form only 3 bonds such that 3 potential bonding sites remain unconnected.

Table 3.

Hydrogen H_2	Oxygen O_2	Nitrogen N_2	Water H_2O	Hydrogen Peroxide H_2O_2

Ammonia NH_3	Methane CH_4	Dichloromethane CH_2Cl_2	Chloroethanol C_2H_5ClO	Carbon Dioxide CO_2

Acetylene C_2H_2	Ethanol C_2H_6O	Acetic Acid $C_2H_4O_2$	Benzene C_6H_6	Iron (III) Oxide Fe_2O_3

Summing Up

1. a. Atoms combine to form molecules in specific ratios. In a water molecule, for example, there are two hydrogens for every one oxygen. If this ratio were different—say two hydrogens to two oxygens—would the shape of the molecule also be different?

_____

   b. Would you still have a water molecule?

_____

2. a. How are the structures for methane and dichloromethane similar?

_____

_____

   b. How are they different?

_____

_____

3. How many *different* structures (configurations) are possible for the formula C_2H_5ClO? (Hint: there are more than two.)

_____

4. Of the 15 models you made, which are linear?

_____

5. Which molecules have multiple bonds between atoms?

_____

6. Which of your structures is flat like a pancake?

_____

CONCEPTUAL PHYSICAL SCIENCE **EXPLORATIONS**	**Activity**

Molecular Mixtures: Radial Paper Chromatography

Circular Rainbows

Purpose
In this activity, you will separate the different colored components of black ink.

Required Equipment and Supplies
a variety of black felt-tip pens
circular filter paper or chromatography paper
a variety of solvents
    water
    methanol (wood alcohol)
    ethanol (grain alcohol)
    isopropyl alcohol (rubbing alcohol)
    white distilled vinegar
    diethyl acetate (finger nail polish remover)
beakers or crucibles upon which to place the paper

Discussion
Black ink is made by combining many different colored inks such as blue, red, and yellow. Together, these inks serve to absorb all the frequencies of light. With no light reflected, the ink appears black.

Procedure
It is easy to separate the components of an ink using a technique called *paper chromatography*. Place a concentrated dot of the ink at the center of a porous piece of paper, such as filter paper. Prop the marked paper on a small beaker or crucible, then carefully add a drop of solvent such as water, white vinegar, or rubbing alcohol on top of the dot. Watch the ink spread radially with the solvent. Because the various components have differing affinities for the solvent, they travel with the solvent at different rates. Just before your drop of solvent is completely absorbed, add a second drop, then a third, and so on until the components have separated to your satisfaction. How the components separate depends on several factors, including your choice of solvent and your technique. Black felt-tip pens tend to work best, but you should experiment with a variety of different types of pens—even food coloring. It is also interesting to watch the leading edge of the moving ink under a microscope. Check for capillary action!

Summing Up
1. The different components of black ink not only have differing affinities for the solvent, they also have differing affinities for the paper, which is fairly polar. How might you expect a component that is ionically charged to behave while using a relatively nonpolar solvent, such as acetone—would it readily travel with the solvent or might it stay behind?

_____

2. Did you ever notice that not all "blue" inks are the same color blue? How might this be explained?

_____

CONCEPTUAL PHYSICAL SCIENCE **EXPLORATIONS**	**Activity**

Molecular Mixing: Purification of Brown Sugar

Pure Sweetness

Purpose
In this activity, you will isolate white sugar from brown sugar.

Required Equipment and Supplies
brown sugar
kitchen cooking pot or 2000-mL beaker
empty glass ketchup bottle or 1000-mL beaker
kitchen knife or microspatula;
small wide-mouth jar (such as a baby food jar)
safety goggles
gloves

Discussion
A supersaturated solution of brown sugar in water may be prepared by dissolving ample amounts of brown sugar in a small amount of boiling water. Upon cooling and after several days, a large number of crystals will have grown. These crystals will still be brown but not quite as brown as the original brown sugar. Recrystallization of these brown crystals will result in crystals that are even less brown. Repeated recrystallization may result in crystals that appear white.

The formation of crystals is almost as much of an art as it is a science. It is difficult, therefore, to guarantee that a particular procedure will always result in crystals of a given size and quality. With this in mind, expect to make some on-the-fly modifications to the following procedure.

This activity may last for several weeks and you should keep a journal of the equipment and materials that you use and the actual procedure that you follow. Be sure to date every entry. Drawing sketches of your set-up is also important.

Procedure
1. In a cooking pot or large beaker, bring about 200 mL of water to a boil. Slowly add brown sugar to the boiling water with continuous stirring. Continue adding brown sugar until it becomes thick and begins to froth.

2. Allow the solution to cool before pouring it into an empty glass ketchup bottle or into a 1000-mL beaker. Allow the syrup to stand for several days until a significant number of crystals have formed. The syrupy solution from which the crystals form is known as the *mother liquor*. Your crystals at this point will likely appear as tiny hard clumps.

3. After there has been a significant amount of crystal growth, pour out the mother liquor. You may discard the mother liquor, but first note its close resemblance to molasses. In fact, that's exactly what it is! The brown color results from the many plant by-products it contains. The mother liquor will pour out very slowly. Consider leaving it inverted for several hours. The crystals should stay behind, stuck to the glass walls.

4. Rinse the collected crystals with warm water to remove additional mother liquor. Rinse only briefly as you want to avoid dissolving these crystals. After this quick rinse, isolate a few of these crystals and examine them closely.

5. Use hot water to help collect all your crystals from your container. Use as little hot water as you can. Transfer your crystals with the small amount of water to a small pot or beaker and apply heat until the crystals dissolve and some frothing is seen. You should end up with a syrupy solution once again.

6. Allow the syrupy solution to cool before transferring it back to the cleaned ketchup bottle or beaker. A glass baby food jar also works well. Wait several days for more crystals to grow. This is called *recrystallization* because you are crystallizing a material that had already been crystallized. The effect is an increase in the crystal's purity.

7. After crystal growth, the mother liquor of this solution can be poured out. The crystals that remain behind can be collected onto a towel, rinsed briefly with warm water and quickly dried so that they retain crystal shape. If crystals still retain significant brown color, they may be recrystallized again using the general procedures given above.

Summing Up

1. How might you prove that commercial grade white sugar still contains small amounts of molasses?

_____

_____

2. Which is more *pure:* white sugar or brown sugar?

_____

_____

3. Which is more *natural:* white sugar or brown sugar?

_____

_____

4. Write a statement regarding the quality verses quantity of the sugar crystals you obtained from this activity.

_____

_____

CONCEPTUAL PHYSICAL SCIENCE **EXPLORATIONS** | **Activity**

Acids and Bases: Red Cabbage Juice pH Indicator

Sensing pH

Purpose
In this activity, you will isolate and use the pH-sensitive pigment of red cabbage.

Requires Equipment and Supplies
red cabbage
cooking pot
strainer
clear plastic cups or 100-mL beakers (5)
safety goggles
water
household solutions: white vinegar, ammonia cleanser, grapefruit juice, club soda, toilet bowl cleaner
household powders: baking soda, washing soda, borax, salt
9-volt battery

Discussion
The pH of a solution can be approximated by way of a pH indicator—any chemical whose color changes with pH. Many pH indicators can be found in plants. Red cabbage is one such plant.

> ### Safety!
> In this activity you will be creating a broth of red cabbage in boiling water. Protect your eyes from the splattering of any liquid or the shattering of any glassware by wearing safety goggles. Protect your skin from scalding by handling the boiling water with care. If you work with toilet bowl cleaner, exercise extra caution as this material is very acidic.

Procedure
Step 1: Shred about a quarter of a head of red cabbage. Boil the shredded cabbage in about 500 mL of water for about 5 minutes. Strain the cabbage while collecting the broth, which contains the pH indicator. Allow the broth to cool so that it can be safely handled.

Step 2: Fill five or more clear plastic cups or 100-mL glass beakers halfway with the broth. To each cup add a small amount of a household solution (about 10 mL) or powder (about a teaspoon). Examples of solutions you might use include white vinegar, ammonia cleanser, grapefruit juice, club soda, and toilet bowl cleaner. Examples of powders you might use include baking soda, washing soda, borax, and salt.

Step 3: Note the changes in color and estimate the pH of the solution based upon the following:

Color of red cabbage indicator	dark red	pinkish red	light purple	light green	dark green
pH range	1 - 4	4 - 6.5	6.5 - 7.2	7.2 - 8	8 - 10

Enter the identity of the material you are testing and the estimated pH of the solution it makes in Table 1.

Step 4: Add some fresh broth to a clear cup and carefully submerge the terminals of a 9-volt battery into this broth. Hold onto the battery. Do not drop it into the broth. Look carefully for any change in color around either of the terminals. If you don't see color changes, you may have to dilute the solution. Look also for the formation of bubbles. Dry the 9-volt battery thoroughly after use and return it to your teacher.

Table 1

Household solution or powder	Estimated pH

Summing Up
1. What color changes did you see upon submerging the terminals of the 9-volt battery into the red cabbage broth?

_____

2. Hydroxide ions, OH⁻, are one of the products that form around the terminals of the 9-volt battery dipped into solution. At which terminal (the positive or negative) did these hydroxide ions form?

_____

3. At what terminal on the 9-volt battery (the positive or negative) did you see the formation of bubbles?

_____

CONCEPTUAL PHYSICAL SCIENCE EXPLORATIONS | **Experiment**

Acids and Bases: Titration

Upset Stomach

Purpose
In this experiment, you will measure and compare the acid-neutralizing strengths of antacids.

Required Equipment and Supplies

Equipment
buret stand with clamps
three 250-mL Erlenmeyer flasks
safety goggles
gloves
balance
mortar and pestle
weigh dishes
plastic pipets
well-plates

Chemicals
0.50 *M* HCl solution
0.50 *M* NaOH solution
phenolphthalein indicator
various brands of antacid tablets

Safety!
This is a fairly safe lab. Eye protection, however, must be worn at all times when you or your lab partners are working with solutions of hydrochloric acid, HCl, or sodium hydroxide, NaOH. You may also want to wear gloves to protect your skin, which may sting if it comes in contact with the hydrochloric acid or feel slippery if in contact with the sodium hydroxide. If either of these solutions get onto your skin, tell your teacher and proceed to rinse your skin thoroughly with running water.

Discussion
Overindulging in food or drink can lead to acid indigestion, a discomforting ailment that results from the excess excretion of hydrochloric acid, HCl, by the stomach lining. An immediate remedy is an over-the-counter antacid, which consists of a base that can neutralize stomach acid. In this experiment, you will add an antacid to a simulated upset stomach. Not all the acid will be neutralized, however, and so you will then determine the effectiveness of the antacid by determining the amount of acid that remains.

This is done by completing the neutralization with another base, sodium hydroxide, NaOH. The reaction between hydrochloric acid and sodium hydroxide is

HCl	**+**	**NaOH**	$\rightarrow$	**NaCl**	**+**	**H$_2$O**
hydrochloric acid		sodium hydroxide		salt		water

The concentrations of the HCl and NaOH used in this experiment are the same. This means that the volume of NaOH needed to neutralize the remaining HCl in the "relieved" stomach is equal to the volume of HCl *not* neutralized by the antacid.

Procedure

Part A: Preparing the Upset Stomach

Step 1: Obtain a 250-mL Erlenmeyer flask.

Step 2: Have your teacher or another authorized individual deliver about 30.00 mL of a 0.50 M HCl solution to your 250-mL Erlenmeyer flask.

Step 3: With the proper number of decimal places as specified by your teacher, record in your data sheet the volume of 0.50 M HCl that your flask actually contains . For example, depending on the instrument used to deliver the HCl solution, your volume might be 29.93 mL or 29.9 mL or simply 30 mL.

Step 4: Add 2 drops of phenolphthalein indicator to the flask. No color change should be observed.

Part B: Adding the Antacid

Step 1: Record the brand and active ingredient of your antacid on the data table.

Step 2: Crush and grind the antacid tablet with a mortar and pestle. (**Note:** Alka-Seltzer need not be crushed.)

Step 3: Carefully transfer all of the resulting powder to a weigh-dish, and then determine and record the mass of the weigh-dish plus powder.

Step 4: Carefully transfer the antacid from the weigh-dish to the upset stomach flask prepared in Part A, swirling for a few minutes while being careful not to spill. This flask now represents your "relieved upset stomach." (The solution should remain colorless.)

Step 5: Determine and record the mass of the empty weigh-dish.

Step 6: Calculate the mass of the antacid tablet.

Part C: Completing the Neutralization

Step 1: Have your teacher show you the proper technique for delivering drops of liquid from a plastic pipet. This involves holding the pipet firmly while gently squeezing the bulb so that you can control drops coming out of the pipet one at a time. Practice your technique by delivering drops of water over a sink.

Step 2: Use a clean pipet to transfer 20 drops of the solution in the "relieved upset stomach" flask to one of the wells. Keep a careful count of the number of drops added, and add this solution until a light pink color persists for at least 30 seconds.

Step 3: Record the number of drops in the data table, and then repeat step 2 for the remaining wells.

Step 4: Repeat the entire procedure, starting from Part A, for two other brands of antacid.

Data

Parts A and B: Preparing the Upset Stomach and Adding the Antacid

	Antacid 1	Antacid 2	Antacid 3
1. Antacid Used	_____	_____	_____
2. Active Ingredient	_____	_____	_____
3. Volume of HCl added to Erlenmeyer flask	_____ mL	_____ mL	_____ mL
4. Number of drops of phenolphthalein added	_____ drops	_____ drops	_____ drops
5. Mass of weigh-dish with crushed antacid	_____ g	_____ g	_____ g
6. Mass of weigh-dish after antacid was transferred	_____ g	_____ g	_____ g
7. Mass of antacid added to "stomach"	_____ g	_____ g	_____ g

Part C: Completing the Neutralization

Antacid 1	Well 1	Well 2	Well 3	Well 4	Well 5		Total
1. Drops of relieved stomach fluid	20	20	20	20	20	=	100
2. Drops of NaOH added to complete neutralization						=	
3. Drops of stomach acid neutralized by NaOH						=	
4. Drops of stomach acid neutralized by antacid						=	

Antacid 2	Well 1	Well 2	Well 3	Well 4	Well 5		Total
1. Drops of relieved stomach fluid	20	20	20	20	20	=	100
2. Drops of NaOH added to complete neutralization						=	
3. Drops of stomach acid neutralized by NaOH						=	
4. Drops of stomach acid neutralized by antacid						=	

Antacid 3	Well 1	Well 2	Well 3	Well 4	Well 5		Total
1. Drops of relieved stomach fluid	20	20	20	20	20	=	100
2. Drops of NaOH added to complete neutralization						=	
3. Drops of stomach acid neutralized by NaOH						=	
4. Drops of stomach acid neutralized by antacid						=	

Summing Up

1. List the antacids you tested in order of neutralizing strength, strongest first:

   strongest _____ > _____ > _____ weakest

2. List the antacids you tested in order of the masses of the tablets, most massive first:

   Most massive _____ > _____ > _____ least massive

3. Divide the total number of drops of stomach acid neutralized by each antacid by the mass of the antacid that was added to the stomach:

   Antacid 1: _____ drops/gram

   Antacid 2: _____ drops/gram

   Antacid 3: _____ drops/gram

4. List the antacids in order of neutralizing strength based upon the number of drops of acid neutralized for every gram of antacid:

   strongest _____ > _____ > _____ weakest

5. What would be the effect on the neutralizing strength for an antacid if the following errors were made?

   a. A student uses two tablets of one of the antacids rather than just one tablet:

   _____

   b. A student spills some of the crushed antacid as it is transferred to the "stomach" flask:

   _____

   c. A student "overshoots" the number of drops of sodium hydroxide so that the solution turns a dark pink rather than a light pink:

   _____

Oxidation and Reduction: Percent Oxygen in Air

Tubular Rust

Purpose

In this activity, you will determine the percent oxygen in air.

Required Equipment and Supplies

steel wool
test tube (15 cm)
pencil
250-mL beaker
centimeter stick and rubber band
white distilled vinegar (5% acetic acid)
ring stand and test tube clamp

Discussion

This activity takes advantage of the rusting of iron by oxygen to determine the percent oxygen in the air. Iron is placed in an air-filled test tube, which is then inverted in water. As the iron reacts with the oxygen, the pressure inside the test tube decreases, and atmospheric pressure pushes water up into the tube. The decrease in volume of air in the test tube is a measure of the depletion of its oxygen content. By measuring the quantity of air in the test tube before and after the rusting of the iron, the percent oxygen in air by volume can be calculated.

Procedure

Step 1: Measure about 1 gram of steel wool that has been pulled apart to increase its surface area. To promote rusting, dip the wool into some white distilled vinegar. Shake off the excess vinegar, and push the steel wool halfway into a test tube 15-cm in length using a pencil. The wool should be packed as loosely as possible to maximize its surface area but packed tight enough so that it doesn't fall out upon inverting the tube.

Step 2: Attach a lightweight centimeter stick to the test tube using a rubber band. Position the stick so that the 0-mm mark is toward the open end. Carefully invert the test tube into the beaker, which should be filled halfway with water. Clamp the test tube to a ring stand and adjust its height so that the lip is no more than a centimeter under the water. Adjust the ruler so that the 0-mm mark is even with the water level *inside* the tube.

Step 3: Water will climb up into the test tube as the rusting proceeds. As this occurs, lower the test tube deeper into the beaker so that the water levels outside and inside the test tube remain even. Begin to read and record the water level inside the test tube at 3-minute intervals until it stops rising.

Step 4: Plot a graph of the water level inside the test tube versus time.

Step 5: Calculate the fraction of oxygen in air as the maximum height of the water inside the tube divided by the total length of the tube. Multiply by 100 to obtain a percentage.

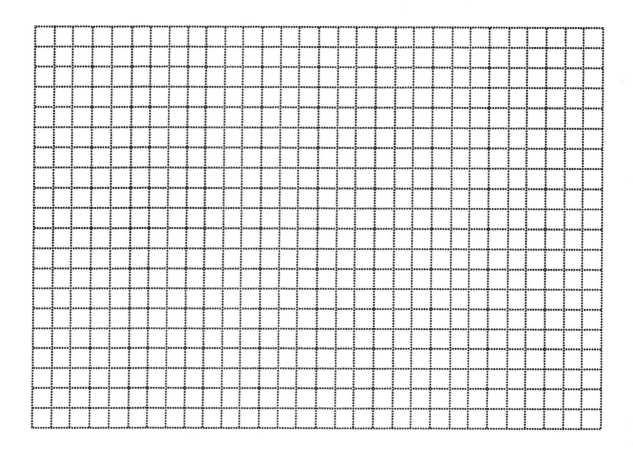

Summing Up

1. What is your experimental percent oxygen in the air?

_____

_____

2. Why is it important that the water levels inside and outside the test tube remain even?

_____

_____

3. Would it take a longer or shorter time for the oxygen to be depleted if the steel wool were packed tightly at the bottom of the test tube? Briefly explain.

_____

_____

CONCEPTUAL PHYSICAL SCIENCE **EXPLORATIONS** | **Activity**

Organic Compounds: Preparation of Fragrant Esters
Smells Great!

Purpose
In this activity, you will prepare a variety of fragrant esters.

Required Equipment and Supplies
watch glasses (5-10 cm diameters)
concentrated sulfuric acid in dropper bottle
safety goggles
various alcohols and carboxylic acids to be reacted together

Alcohols		Carboxylic Acids
methanol	+	salicylic acid
octanol	+	acetic acid
benzyl alcohol	+	acetic acid
isoamyl alcohol	+	acetic acid
n-propanol	+	acetic acid
isopentenol	+	acetic acid
isobutanol	+	propionic acid

Discussion
Many flavorings and fragrances, both natural and artificial, are a class of organic molecules called *esters*. These molecules contain a carbonyl group bonded to an oxygen atom that is bonded to a carbon atom (See Chapter 27, *Conceptual Physical Science: Explorations*). Esters can be prepared in the laboratory by reacting alcohols with carboxylic acids (Figure 1).

Methanol
(an alcohol)   Butyric Acid
(a carboxylic acid)   Methyl Butyrate
(an ester)

Figure 1. An example of the formation of an ester from an alcohol and a carboxylic acid.

In this activity you and your class will prepare small but fragrant quantities of different kinds of esters and attempt to identify their smells. To minimize the production of organic wastes, each lab group may be assigned specific esters to prepare.

SAFETY!
This lab requires the use of concentrated sulfuric acid, H_2SO_4. Only one drop per reaction will be used, but because this acid is concentrated, these drops can be very harmful to skin. Your teacher will be dispensing drops of concentrated sulfuric acid onto your watch plates. Handle your samples with care. They may smell good but do not taste them because they may still contain residual amounts of sulfuric acid. Wear safety goggles at all times. When you are finished with this activity empty your samples into the provided waste container. Rinse your watch glasses with the solvent provided for you by your teacher and then clean them with soap and water.

Procedure

Step 1: Place 1 mL of an alcohol along with 1 mL (or 1 gram if a solid) of the corresponding carboxylic acid as given in Table 1 onto a watch glass. Smell your samples before the addition of sulfuric acid. Note: it is improper and unsafe laboratory practice to stick the sample really close to your nose and sniff heavily. Instead, odors can be brought to the nose by waving your hand over the watch glass, which may be held several inches away from your face.

Step 2: To initiate the formation of an ester from the alcohol and the carboxylic acid, bring your samples to the teacher who will add a drop of concentrated sulfuric acid. Gently stir with a microspatula to promote mixing. As the ester forms, a noticeable change in odor will occur. Some samples may take longer than others—allow up to 10 minutes.

Step 3: Complete Table 1 by matching the observed smell with one of the following fragrances. Check with other lab groups for samples that were not assigned to you.

Possible fragrances include apple, banana, "Juicy Fruit," orange, peach, pear, rum, and wintergreen.

Table 1.

Alcohol	Carboxylic Acid	Observed Smell
methanol	salicylic acid	
octanol	acetic acid	
benzyl alcohol	acetic acid	
isoamyl alcohol	acetic acid	
n-propanol	acetic acid	
isopentenol	acetic acid	
methanol	butyric acid	
isobutanol	propionic acid	

Summing Up

1. What's the difference between an ester found in a natural product, such as a pineapple, and the same ester produced in the laboratory?

2. Why do foods become more odorous at higher temperatures?

Name_____ Period_____ Date_____

Polymers: Densities of Organic Polymers

Name That Recyclable

Procedure
In this activity, you will identify a variety of unknown recyclable plastics based upon their densities.

Required Equipment and Supplies
recyclable plastics (PET, HDPE, LDPE, PP, PS)
solutions
     95% ethanol and water (1:1 by volume)
     95% ethanol and water (10:7 by volume)
     10% NaCl in water

Discussion
There are many different types of plastics and many of them are recyclable. Ultimately plastics need to be sorted according to their composition. For this reason plastics are coded with a number within the recycling arrow logo. The initials of the type of plastic may also appear. For example, the plastic used to make 2-liter soft-drink bottles is polyethylene terephthalate (PET). This plastic has the following recycling code:

PET

Plastics can be also be identified based upon their densities, which is useful if the recycling code is unreadable or absent. In this activity you are to identify pieces of unknown plastics based upon their densities.

Procedure
Step 1: Use the information in Table 1 to develop a separation scheme by which you will be able to identify all your unknown pieces of plastic.

Step 2: Use the solutions provided to identify each unknown piece of plastic:

Unknown No.:					
Identity:					

Table 1.

Recyclable Plastic	Density Compared to...			
	Water	Ethanol and Water (1:1)	Ethanol and Water (10:7)	10 % NaCl in Water
1 **PET** Polyethylene terephthalate	Greater	Greater	Greater	Greater
2 **HDPE** High density polyethylene	Less	Greater	Greater	Less
4 **LDPE** Low density polyethylene	Less	Less	Greater	Less
5 **PP** Polypropylene	Less	Less	Less	Less
6 **PS** Polystyrene	Greater	Greater	Greater	Less

Summing Up

1. What other physical properties might be used to identify unknown pieces of plastic?

_____

_____

2. You've just been given a thousand pounds of recyclable polystyrene and polypropylene plastic pieces all mixed together. Suggest how you might quickly separate the different types of plastic from one another.

_____

_____

110

CONCEPTUAL PHYSICAL SCIENCE EXPLORATIONS	Activity

Minerals and Their Formation

Crystal Growth

Purpose
In this activity, you will observe the growth of crystals from a melt and from a solution.

Required Equipment and Supplies
thymol
sodium chloride
sodium nitrate
potassium aluminum sulfate (alum)
copper acetate
glass slide or clear glass plate
forceps
petri dish
hot plate
microscope

Discussion
A **mineral** is a naturally formed inorganic solid composed of an ordered array of atoms. The atoms in different minerals are arranged in their own characteristic ways. This systematic arrangement of atoms is known as a mineral's crystalline structure, which exists throughout the entire mineral specimen. If crystallization occurs under ideal conditions, it will be expressed in perfect crystal faces.

I. GROWTH OF CRYSTALS FROM A MELT
Most minerals originate from a magma melt. Crystalline minerals form when molten rock cools. The temperature at which minerals crystallize is very high (well above the boiling point of water) making direct observation difficult. In this experiment we will use thymol, an organic chemical that crystallizes near room temperature. Although thymol is different from a magma, the crystallization principles exhibited are similar.

SAFETY!
Thymol is not poisonous but may cause skin and eye irritation. Be safe: use forceps to handle thymol.

Procedure

Part A. Slow Cooling Without "Seed" Crystals
Set the hot plate to low heat. Place a petri dish containing a small amount of crystalline thymol on the hot plate until all the crystals are melted. Allow the melt to heat for one to two minutes and then set it aside to cool slowly. Do not disturb during cooling. This melt will be examined at the end of the lab session.

Part B. Slow Cooling With "Seed" Crystals

Repeat above procedure (A) but transfer the petri dish to the stage of a microscope as soon as the thymol is melted. Add several (four or five) "seed" crystals to the melt and observe it under the microscope. As the melt cools, you will see the crystals begin to grow.

Describe the manner of crystal growth. Think about the rate of growth, the direction of growth, the crystal faces, and the effect of limited space on crystal shape. What role do the "seed" crystals play in initiating crystal growth?

_____

_____

3. Make a sketch of the final crystalline solid. The crystals produced in this experiment are similar to the type of crystals found in common igneous rocks. Both are interlocking crystals. Compare the thymol crystals to those in a rock specimen of granite. Sketch the texture of the granite. How do the thymol crystals compare with those in granite?

_____

_____

Part C. Rapid Cooling

Repeat above procedure (A) but transfer the petri dish to the top of an ice cube and let it cool for 30 seconds. The melt will cool very rapidly. Quickly move the petri dish to the microscope and observe the nature of the crystal growth. Time permitting, repeat this procedure until you are sure of your observations and can state your general conclusions.

_____

_____

Part D. Examination of Crystals From Procedure Part A

Examine the thymol crystals from the first part of this exercise. Observe the crystal size. What effect does the rate of cooling have on the crystal size?

_____

_____

II. GROWTH OF CRYSTALS FROM SOLUTION

Many minerals are precipitated from aqueous solutions by evaporation. Crystal growth of such minerals can be observed in the laboratory by evaporating prepared concentrated solutions.

Procedure

Obtain concentrated solutions of sodium chloride, sodium nitrate, potassium aluminum sulfate (alum), and copper acetate from your instructor.

Place a drop of each solution on a microscope slide. Label each slide accordingly. As the water evaporates from the slide, crystals of each compound will appear. With time and continued evaporation, the crystals will grow larger.

Summing Up

1. How does crystallization from an aqueous solution compare to crystallization from a melt?

_____

_____

2. Do the crystals precipitated from a solution have unique crystal forms?

_____

_____

3. According to your observations in this activity, do you think that crystal forms provide a good reference for mineral identification?

_____

_____

4. According to your observations in this activity, what is the relationship between cooling and crystal size?

_____

_____

Minerals and Their Formation: Identification of Minerals

What's That Mineral?

Purpose

In this activity, you will observe the physical properties of various mineral samples and identify them by a systematic procedure.

Required Equipment and Supplies

mineral collection
hardness set—piece of glass, steel knife, copper penny,
streak plate (non-glazed porcelain plate)
dilute hydrochloric acid (HCl)

Discussion

A mineral is a naturally formed inorganic solid with a characteristic chemical composition and crystalline structure. The different combinations of its elements and arrangement of atoms determine the physical properties of a mineral: its shape, the way it reflects light, its color, hardness, and its mass.

I. The physical properties dependent on a mineral's chemical composition include luster, streak, color, specific gravity, and reaction with acid.

The **luster** of a mineral is the appearance of its surface when it reflects light. Luster is independent of color; minerals of the same color may have different lusters and minerals of the same luster may have different colors. Mineral luster is classified as either metallic or nonmetallic.

TEST FOR LUSTER	
Metallic minerals are usually	**Nonmetallic minerals are usually**
a. gold, silver or black	a. not gold or silver
b. shiny, polished	b. shiny to dull
c. opaque	c. transparent, translucent, or opaque
d. always have a streak	d. rarely have a streak

The **streak** of a mineral is the color of its powdered form. We can see a mineral's streak by rubbing it across a nonglazed porcelain plate. Minerals with a metallic luster generally leave a dark streak that may be different from the color of the mineral. For example, the mineral hematite is normally reddish-brown to black, but always streaks red. Magnetite is normally iron-black, but streaks black. Limonite is normally yellowish-brown to dark brown, but always streaks yellowish-brown. Minerals with a nonmetallic luster either leave a light streak or no streak at all.

TEST FOR STREAK
1. Scrape the mineral across a non-glazed porcelain plate.
2. Blow away excess powder.
3. The color of the powder is the streak.

Although **color** is an obvious feature of a mineral, it is not a reliable means of identification. When used with luster and streak, color can sometimes aid in the identification of metallic minerals. Nonmetallic minerals may occur in a variety of colors or be colorless. Therefore, color is not used for the identification of nonmetallic minerals. In this exercise, we will only differentiate between light-colored and dark-colored minerals.

Specific gravity, (s.g.), is the ratio of the mass (or weight) of a substance to the mass (or weight) of an equal volume of water. Metallic minerals tend to have a higher specific gravity than nonmetallic minerals. For example, the metallic mineral gold (Au) has a specific gravity of 19.3 whereas quartz (SiO_2), a nonmetallic mineral, has a specific gravity of 2.65.

The **reaction to acid** is an important chemical property often used to identify carbonate minerals. Carbonate minerals effervesce (fizz) in dilute hydrochloric acid (HCl). Some carbonate minerals react more readily with HCl than others. For instance, calcite ($CaCO_3$) strongly effervesces when exposed to HCl, but dolomite ($CaMg(CO_3)_2$) doesn't react unless it is scratched and powdered.

II. The physical properties dependent on a mineral's crystalline structure include hardness, cleavage and fracture, crystal form, striations, and magnetism.

The resistance of a mineral to being scratched (or its ability to scratch) is a measure of the mineral's **hardness**. The varying degrees of hardness are represented on Mohs scale of hardness. For this activity, we are concerned with the hardness of some common objects.

Cleavage and fracture are useful guides for identifying minerals. **Cleavage** is the tendency of a mineral to break along planes of weakness. Planes of weakness depend on crystal structure and symmetry. Some minerals have distinct cleavage. Mica, for example, has perfect cleavage in one direction, and breaks apart to form thin, flat sheets. Calcite has perfect cleavage in three directions and breaks to produce rhombohedral faces that intersect at 75-degrees. A break other than along cleavage planes is a **fracture**.

MOHS SCALE OF HARDNESS		
Mineral	**Scale Number**	**Common Objects**
Diamond	10	
Corundum	9	
Topaz	8	
Quartz	7	Steel file
Feldspar	6	Window glass
Apatite	5	Pocket knife
Fluorite	4	
Calcite	3	Copper penny
Gypsum	2	Fingernail
Talc	1	

TEST FOR HARDNESS
1. Place a glass plate on a hard flat surface.
2. Scrape the mineral across the glass plate.
3. If the glass is scratched, the mineral is harder than glass.
4. If the glass is not scratched, the mineral is softer than glass.
5. If the mineral is softer than glass, test to see if it is harder than a copper penny. Scrape the mineral across the penny.
6. If the mineral is softer than the penny, test to see if it is harder than your fingernail. Try to scratch the mineral with your fingernail.

Every mineral has its own characteristic **crystal form**. Some minerals have such a unique crystal form that identification is relatively easy. The mineral pyrite, for example, commonly forms as intergrown cubes, while quartz commonly forms as six-sided prisms that terminate in a point. Most minerals, however, do not exhibit their characteristic crystal form. Perfect crystals are rare in nature because minerals typically grow in cramped, confined spaces.

CLEAVAGE PATTERNS			
Number of Cleavage Directions	Shape	Number of Flat Surfaces	Sample Illustration
1	Flat sheets	2	
2 at 90°	Rectangular cross-section	4	
2 not at 90°	Parallelogram cross-section	4	
3 at 90°	Cube	6	
3 not at 90°	Rhombohedron	6	
Fracture	Irregular shape	0	

Some minerals have grooves on their cleavage planes. These grooves, called **striations**, can be used to differentiate between feldspar minerals. Plagioclase feldspars have straight parallel striations on one cleavage plane. Orthoclase feldspars have lines that resemble striations but are actually color variations within the mineral. These grooves are not straight, and they are not parallel to each another. Instead, these "striations" make a criss-cross pattern.

Some minerals exhibit magnetism. To test for magnetism, simply expose the mineral to a small magnet or a compass.

So we see that physical properties that depend on a mineral's crystalline structure include hardness, crystal form, cleavage and fracture, striations, and magnetism.

MINERALS WITH METALLIC LUSTER

Streak Color	Properties	Mineral
Black to gray	silver gray 3 directions of cleavage at 90° cubic crystals. hardness = 2.5 specific gravity = 7.6	Galena
Black to gray	black to dark gray magnetic hardness = 6 specific gravity = 5.	Magnetite
Black to gray	gray to black marks paper feels greasy hardness = 1-2 specific gravity = 2.2 golden yellow may tarnish purple hardness = 4	Graphite Chalcopyrite
Black to greenish black	specific gravity = 4.2 brass yellow may tarnish green cubic crystals striations hardness = 6 specific gravity = 5.2	Pyrite (fool's gold)
Reddish brown to black	silver to gray may tarnish reddish brown hardness = 5-6 specific gravity = 5	Hematite
Yellowish brown to reddish brown	black to silver gray to golden brown may tarnish yellowish brown hardness = 5.5 specific gravity = 4	Limonite

Some are interesting...
some are more interesting!

118

Minerals with Nonmetallic Luster—Dark Colored

Hardness	Cleavage	Properties	Mineral
Harder than glass	Present	black to blue gray 2 directions of cleavage, not quite at 90° striations on one cleavage plane hardness = 6 specific gravity = 2.7	Plagioclase
		dark green to black 2 directions of cleavage at nearly 90° no striations hardness = 6 specific gravity = 3.5	Pyroxene
		dark green to black 2 directions of cleavage intersecting at 60° and 120° no striations hexagonal crystals hardness 5.5 specific gravity = 3.6	Amphibole (hornblende)
Softer than glass		brown to dark green to black I direction of cleavage producing thin sheets translucent hardness = 3 specific gravity	Biotite
		green to dark green 1 direction of cleavage producing thin curved sheets greasy luster hardness = 2.5 specific gravity = 2.9	Chlorite

Earth science is down to earth!

119

Minerals with Nonmetallic Luster—Dark Colored

Hardness	Cleavage	Properties	Mineral
Harder than glass	Absent	olive green to black no streak glassy luster conchoidal fracture hardness 7 specific gravity = 2.6	Olivine
		light to dark gray no streak glassy, transparent, translucent conchoidal fracture hexagonal crystals hardness = 7 specific gravity = 2.6	Quartz
		deep red to brown no streak translucent conchoidal fracture isometric crystals hardness = 7 specific gravity = 4	Garnet
Variable		red reddish brown streak opaque, earthy uneven fracture hardness = 1.5 - 5.5 specific gravity = 5.2	Hematite

When I dream of geology
I have rocks in my head!

Minerals with Nonmetallic Luster—Light Colored

Hardness	Cleavage	Properties	Mineral
Harder than glass	Present	pale orange-pink, white green, brown 2 cleavage directions at nearly 90° no striations color lines on cleavage planes hardness = 6 specific gravity = 2.6	Orthoclase feldspar
		white to blue-gray 2 directions of cleavage not quite at 90° striations on one cleavage plane hardness = 6 specific gravity = 2.7	Plagioclase feldspar
Softer than glass		colorless to white 3 directions of cleavage at 90° soluble in water hardness = 2.5 specific gravity	Halite
		colorless to white I direction of cleavage hardness = 2 (easily scratched with fingernail) specific gravity = 2.3	Gypsum
		colorless to white or yellow 3 directions of cleavage not at 90° (rhomb shaped) translucent to transparent strong reaction to acid hardness = 3 specific gravity = 2.7	Calcite

Hmmm...gold maybe?

Minerals with Nonmetallic Luster—Light Colored

Hardness	Cleavage	Properties	Mineral
Softer than glass	Present	white, gray, pink 3 directions of cleavage not at 90° (rhomb shaped) opaque reacts to acid when powdered hardness = 3.5 specific gravity = 2.9	Dolomite
		yellow, blue, green, violet 4 directions of cleavage transparent to translucent cubic crystals hardness = 4 specific gravity = 3.2	Fluorite
		colorless to pale green I direction of cleavage producing thin elastic sheets hardness = 2.5 specific gravity = 2.7	Muscovite
		white to greenish I direction of cleavage pearly luster hardness = I specific gravity = 2.8	Tale
Harder than glass	Absent	white, gray, pink, violet glassy luster conchoidal fracture hardness = 7 specific gravity = 2.6	Quartz
		olive green to yellow green glassy luster conchoidal fracture hardness = 7 specific gravity = 3.5	Olivine

Minerals...sigh!

Procedure

Examine the various unknown minerals and note their characteristics on the following worksheet. Identify the different minerals by comparing your list to the mineral identification tables.

Separate metallic and nonmetallic minerals
    A. If mineral is metallic determine:
        1. streak
        2. color
        3. hardness
        4. any other distinguishing properties
        5. the name of the mineral

    B. If the mineral is nonmetallic determine:
        1. color (separate light minerals from dark minerals)
        2. hardness
        3. cleavage
        4. any other distinguishing properties
        5. the name of the mineral

Mineral Identification Sheet

Luster	Streak	Color	Hardness	Cleavage	Other Characteristics	Mineral Name

Summing Up

1. What distinguishing characteristic is used in identifying the following minerals?

   a) halite _____

   b) pyrite _____

   c) quartz _____

   d) biotite _____

   e) fluorite _____

   f) garnet _____

2. Indicate whether the following physical properties result from a mineral's crystalline structure or a mineral's chemical composition.

   a) crystal form _____

   b) color _____

   c) cleavage _____

   d) specific gravity _____

3. If a mineral does not exhibit a streak, is it metallic? Explain.

   _____

   _____

4. Which property is more reliable in mineral identification, color or streak? Why?

   _____

   _____

5. What physical properties distinguish biotite from muscovite?

   _____

   _____

6. What physical properties distinguish plagioclase feldspars from orthoclase feldspars?

   _____

   _____

CONCEPTUAL PHYSICAL SCIENCE **EXPLORATIONS**	**Activity**

Rocks

Rock Hunt

Purpose
In this activity, you will practice finding the geology all around you.

Required Equipment and Supplies
desire

Discussion and Procedure

Activity 1
Go on a rock hunt. Now that you are more familiar with rocks and minerals you can start your own collection. Gather rocks from the beach, an old river channel, a stream bed, a road cut, or even your back yard. Collect at least six different looking rocks and classify them into the three major rock groups. What tell-tale features help you classify the rocks? Can you identify your rocks by these features?

Activity 2
Rocks are not only found on the beach and in the mountains, but almost everywhere. In fact, if you live in a city you are probably surrounded by more rock materials than you realize. Take a field trip down any city street and you will notice that most buildings are constructed from stone material. Many stone buildings are polished, providing an easy view of a rock's mineral composition, texture, and hence its method of formation. Is the rock composed of visible crystals? Are the crystals interlocking? Are the crystals flattened or deformed? What is the grain size? Are there fossils? Support your classification with your observations.

Activity 3
Observe the buildings in your locality. Look for the older buildings, most of which were built from local material. The different rock types used in the construction of some of these buildings can tell you much about local history. Is there a marble or granite quarry in your area? Try to trace the building material to its origin.

CONCEPTUAL PHYSICAL SCIENCE **EXPLORATIONS**	**Activity**

Rocks: Identification of Rocks

What's That Rock?

Purpose
In this activity, you will identify rocks from the three different types: igneous, metamorphic, and sedimentary.

Required Equipment and Supplies
collection of assorted igneous, sedimentary, and metamorphic rocks
dilute hydrochloric acid (HCl)

Discussion
The three major classes of rocks—igneous, sedimentary, and metamorphic—have their own distinct physical characteristics. By learning to identify the representative characteristics of each rock type we may gain a better understanding of the history recorded in the earth's crust.

Procedure
You will be given three sets of rocks—igneous, sedimentary, and metamorphic. Your task is to identify the rocks in each set. The first set to identify consists of igneous rocks. Refer to Table 1, *Classification of Igneous Rocks* to aid in your identification. The second set to identify consists of sedimentary rocks. For this part of the exercise refer to Table 2, *Classification of Sedimentary Rocks.* To make identification and classification easier, the table has been divided into two parts—Part A for clastic sedimentary rocks, and Part B for nonclastic sedimentary rocks. The final set to identify consists of metamorphic rocks. Refer to Table 3, *Classification of Metamorphic Rocks* to help distinguish the different metamorphic rocks.

Part A: Identification of Igneous Rocks
Molten magma welling up from within the earth produces igneous rock of two types—*intrusive* and *extrusive*. Intrusive rocks are formed from magma that solidified below the earth's surface and extrusive rocks are formed from magma erupted at the earth's surface.

Step 1: The first step in identifying igneous rocks is to observe their texture. Texture is related to the cooling rate of magma in a rock's formation. Magma that solidifies below the earth's surface cools slowly, forming large, visible interlocking crystals that can be identified with the unaided eye. This coarse-grained texture is described as **phaneritic**. If the texture is exceptionally coarse-grained (visible minerals larger than your thumb) the rock is described as having a **pegmatitic** texture. Some intrusive rocks contain two distinctly different crystal sizes in which some minerals are conspicuously larger than the other minerals. This texture, with big crystals in a finer groundmass, is called **porphyritic.**

In contrast, magma that reaches the surface tends to cool rapidly forming very fine-grained rocks. This fine-grained texture is described as **aphanitic**. When magma is cooled so quickly that there is not time for the atoms to form crystals, the texture is described as **glassy**. Many aphanitic rocks contain cavities left by gases escaping from the rapidly cooling magma. These gas bubble cavities are called vesicles, and the rocks that contain them are said to have a **vesicular** texture. If the gaseous magma cools very quickly the texture that develops is described as **frothy** (foam-like glass). In a volcanic eruption, the forceful escape of gases causes rock fragments to be torn from the sides of the volcanic vent. These rock fragments combine with volcanic ash and cinders to produce a **pyroclastic**, or fragmental texture.

CLASSIFICATION OF IGNEOUS ROCKS

		Composition			
		Light-Colored	**Intermediate-Color**	**Dark-Colored**	**Very Dark Color**
		10-20% quartz K-feldspar> plagioclase ≈10% ferromagnesian minerals	No quartz plagioclase> K-feldspar 25-40% ferromagnesian minerals	No quartz plagioclase ≈ 50% 50% ferromagnesian minerals	100% ferromagnesian minerals
Texture	Pegmatitic (very coarse-grained)	Pegmatite			
	Porphyritic (mixed crystal sizes)	Porphyry			
	Phaneritic (coarse-grained)	Granite Granodiorite	Diorite	Gabbro	
	Aphanitic (fine-grained)	Rhyolite Peridotite	Andesite	Basalt	Periodite Dunite
	Glassy	Obsidian			
	Porous (glassy, frothy)	Pumice		Scoria	
	Pyroclastic (fragmental)	Volcanic tuff (fragments < 4mm) Volcanic breccia (fragments > 4mm)			

Table 1

Step 2: Observe the color, and hence the chemical composition of an igneous rock sample. Igneous rocks are either rich in silicon or poor in silicon. Silicon-rich (quartz-rich) rocks tend to be light in color whereas silicon-poor rocks tend to be dark in color. Igneous rocks can therefore be divided into a light-colored group—quartz, feldspars, and muscovite; and a dark-colored group—the ferromagnesian minerals—biotite, pyroxene, hornblende, and olivine. (*Ferromagnesian* refers to minerals that contain iron [ferro] and magnesium [magnesian]).

Part B: Identification of Sedimentary Rocks

Rock material that has been weathered, subjected to erosion, and eventually consolidated into new rock is sedimentary rock. Sedimentary rocks are classified into two types—clastic and nonclastic. **Clastic** sedimentary rocks are formed from the compaction and cementation of fragmented rocks. **Nonclastic** sedimentary rocks are formed from the precipitation of minerals in a solution. This chemical process can occur directly, as a result of inorganic processes, or indirectly, as a result of biochemical reaction.

Step 1: Observe the texture of the rock. If the rock is composed of visible particle grains the rock is probably clastic. Clastic sedimentary rocks are classified according to particle size. Large sized particles range from boulders to cobbles to pebbles. If the large-sized particles are rounded, the rock is a **conglomerate**. If they are angular and uneven, the rock is a **breccia**. Medium sized particles produce the various types of sandstone. **Quartz sandstone** is composed wholly of well rounded and well sorted quartz grains. **Arkose** is composed of quartz with about 25 percent feldspar grains. **Graywacke** is composed of both quartz and feldspar grains with fragments of broken rock in a clay type matrix. Silts and clays comprise the fine sized particles. Silt sized particles produce **siltstone** and clay sized particles produce **shale**.

128

Step 2: If the rock does not fit into the clastic classification, the rock is probably nonclastic. Nonclastic sedimentary rocks are classified by their chemical composition. Inorganic processes produce the **evaporites**—minerals formed by the evaporation of saline water. The evaporites include **rock salt** and **gypsum**. Rock salt can be identified by its salty taste. Biochemical reactions produce **carbonates** by way of calcareous-secreting organisms, and **chert**, by silica-secreting organisms. **Limestone, chalk,** and **dolostone** are carbonate rocks. Recall that carbonate minerals react (fizz) to HCL. Chert can be identified by its waxy luster and conchoidal fracture. One rock, **coal** (a biogenic sedimentary rock) is formed from the accumulation and compaction of vegetation matter.

CLASSIFICATION OF CLASTIC SEDIMENTARY ROCKS

Particle size		Rock Name	Characteristics
Coarse (> 2mm)		Conglomerate	Rounded grain fragments
Coarse (> 2mm)		Breccia	Angular grain fragments
Medium (1/16 - 2mm)	Sandstone	Quartz	Quartz grains (often well rounded, well sorted)
Medium (1/16 - 2mm)	Sandstone	Arkose	Quartz and feldspar grains (often reddish color)
Medium (1/16 - 2mm)	Sandstone	Graywacke	Quartz grains, Small rock fragments, and clay minerals (often grayish color)
Fine (1/256 - 1/16 mm)		Siltstone	Silt sized particles, surface is slightly gritty
Fine (<1/256 mm)		Shale	Clay sized particles, surface has smooth feel, no grit

CLASSIFICATION OF NON-CLASTIC SEDIMENTARY ROCKS

Composition		Rock Name	Characteristics
Halite (NaCl)	Evaporites	Rock Salt	Salty taste
Gypsum ($CaSO_4*2H_2O$)	Evaporites	Gypsum	Inorganic precipitate
Organic matter		Coal	Compacted, carbonized plant remains
Calcium Carbonate ($CaCO_3$)		Limestone	Fine-grained, strong reaction to HCl
Calcium Carbonate ($CaCO_3$)		Chalk	Microfossils; fine-grained
Calcium Carbonate ($CaCO_3$)		Fossiliferous limestone	Macrofossils and fossil fragments
Dolomite ($CaMg(CO_3)_2$)		Dolostone	Reaction to HCl when in powdered form
Quartz (SiO_2)		Chert	Hard, dense, waxy luster, may show conchoidal fracture

Table 2

Part C: Identification of Metamorphic Rocks

Rock material that has been changed in form by high temperature or pressure is metamorphic rock. Metamorphic rocks are most easily classified and identified by their texture; and when a particular mineral is very obvious, by their mineralogy. Metamorphic rocks can be divided into two groups: **foliated** and **nonfoliated**.

Foliated metamorphic rocks have a directional texture and a layered appearance. The most common foliated metamorphic rocks are slate, phyllite, schist, and gneiss

Slate is very fine-grained and is composed of minute mica flakes. The most noteworthy characteristic of slate is its excellent rock cleavage—the property by which a rock breaks into plate-like fragments along flat planes. **Phyllite** is composed of very fine crystals of either muscovite or chlorite that are larger than those in slate, but not large enough to be clearly identified. Phyllite is distinguished from slate by its glossy sheen. **Schists** have a very distinctive texture with a parallel arrangement of the sheet structured minerals (mica, chlorite, and/or biotite). The minerals in a schist are often large enough to be easily identified with the naked eye. Because of this, schists are often named according to the major minerals in the rock (biotite schist, staurolite-garnet schist, etc.). **Gneiss** contains mostly granular, rather than platy minerals. The most common minerals found in gneiss are quartz and feldspar. The foliation in this case is due to the segregation of light and dark minerals rather than alignment of platy minerals. Gneiss has a composition very similar to granite and is often derived from granite.

Nonfoliated rocks are mono-mineralic and thus lack any directional texture. Their texture can be described as coarsely crystalline. Common nonfoliated metamorphic rocks are **marble** and **quartzite**.

CLASSIFICATION OF METAMORPHIC ROCKS

Foliated Metamorphic Rock

Crystal Size	Rock Marne		Characteristics
Very fine, crystals not visible	Slate		Excellent rock cleavage
Fine grains, crystals not visible	Phyllite		Well developed foliation; glossy sheen
Coarse texture Crystals visible with unaided eye Micaceous minerals Often contains large crystals	Schist	Muscovite schist	Mineral content reflects increasing metamorphism from top to bottom
		Chlorite schist	
		Biotite schist	
		Garnet schist	
		Staurolite schist	
		Kyanite schist	
		Sillimanite schist	
Coarse	Gneiss		Banding of light sod dark minerals

Nonfoliated Metamorphic Rock

Precursor Rock	Rock Name	Characteristics
Quartz sandstone	Quartzite	Interlocking quartz grains
Limestone	Marble	Interlocking calcite grains

Table 3

130

Summing Up

1. Did you find evidence that igneous rocks can exhibit both fine and coarse textures? Describe the textures.

_____

_____

_____

_____

2. What are the influencing factors for large crystal formation? What types of rock exhibit enlarged crystals?

_____

_____

_____

3. What distinguishing characteristics are exhibited in sedimentary rocks?

_____

_____

_____

4. How can we distinguish igneous rocks from metamorphic rocks?

_____

_____

_____

Going Further

For this exercise you will first determine if the rock is igneous, sedimentary, or metamorphic. Then you will use the classification tables for the different rock types to identify the rocks. Examine your rock specimen closely. Look at its texture. Can you see individual mineral grains? If so, the texture is coarse to medium coarse. If the mineral grains are too small to be identified, the texture is fine. How are the grains arranged? Are the grains interlocking? Interlocking grains are formed by crystallization, so the rock is probably either igneous or metamorphic. Are the interlocking grains aligned? Do they show foliation? Foliation indicates metamorphism. Are the grains separated by irregular spaces filled with cementing material? Are fossils present? Does the rock react with acid? If so, the rock is sedimentary. If the rock is crystalline and shows no reaction to acid, check the hardness of the rock. Metamorphic and igneous rocks are harder than sedimentary rocks.

Architecture of the Earth: Topographic Maps

Top This

Purpose

In this activity, you will interpret maps, particularly topographic maps, which show landforms and approximate elevations above sea level. Using points of known elevation, you will learn to draw contour lines. Using a topographic map, you will learn to construct a topographic profile.

Required Equipment and Supplies

topographic quadrangle map (provided by your instructor)
ruler
pencil and eraser

Discussion

Maps provide a representation of the earth's surface and are a very useful tool. A map is a scaled down, idealized representation of the real world. Everything on a map must be proportionally smaller than what it really is. Roads, waterways, mountains, ground area, and distance are all proportionally reduced in scale. A map's **scale** is defined as the relationship between distance on a map and distance on the ground. There are three ways to describe scale on a map. A *graphical scale* is a drawing, a line marked off with distance units. A *verbal scale* uses words to describe distance units—"one inch to one mile". The third way is a *representative fraction* (rf) that gives the proportion as a fraction or ratio, such as 1:24,000. This means that one unit of measure on the map—1 inch or 1 centimeter—is equal to 24,000 units of the same measure on the ground. If the scale is 1:10,000 then 1 inch on the map is equal to 10,000 inches on the ground. The first number refers to the map distance and is always 1. The second number refers to the ground distance and will change depending on scale.

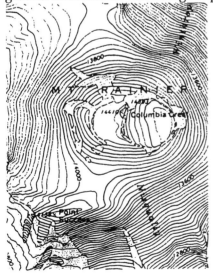  

1:24,0000 scale	1:100,000 scale	1:125,000 scale

Figure 1 U.S. Geological Survey Maps of the same area at different scales. The scale of a map tells us how much area is being represented and the level of detail in that area. When comparing maps, we see that the smaller the scale the larger the area represented, and the larger the scale the smaller the area represented.

133

Large-Scale Map Example: 1:10,000	Small-Scale Map Example: 1:1,000,000
Shows a small area with a lot of detail. A large scale is good for urban, street or hiking maps that require detail.	Shows a large area with very little detail. A small scale is good for world or regional maps that cover a large area.

Scale Conversion

It is difficult to envision 10,000 inches, let alone 1,000,000 inches on the ground. In order to make these units more meaningful we can convert them into miles or kilometers.

Question   One inch on a 1:25,000 scale represents what distance on the ground?

Answer   Using conversion   1 foot = 12 inches
5280 feet = 1 mile
1 mile = 1.609 kilometers

$$25,000 \text{ in} \times \frac{1 \text{ ft}}{12 \text{ in}} \times \frac{1 \text{ mile}}{5280 \text{ ft}} = .394 \text{ miles}$$

$$.394 \text{ miles} \times \frac{1.609 \text{ km}}{1 \text{ mile}} = .634 \text{ kilometers}$$

Question   One mile on the ground is represented by what distance (in inches) on a 1:25,000 scale map?

Answer $$1 \text{ mi} \times \frac{5280 \text{ ft}}{1 \text{ mi}} \times \frac{12 \text{ in}}{1 \text{ ft}} \times \frac{1 \text{ in on map}}{25000 \text{ in}} = 2.53 \text{ inches on map}$$

Relief Portrayal and Topographic Maps

A **topographic map** is a two-dimensional representation of a three-dimensional land surface. Topographic maps show **relief**—the extent to which an area is flat or hilly. To show land surface form and vertical relief, topographic maps use contours. A **contour** line connects all points on the map having the same elevation above sea level. Each contour line acts to separate the areas above that elevation from the areas below it (Figures 2 and 3).

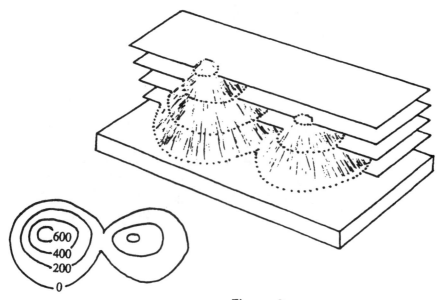

Figure 2

134

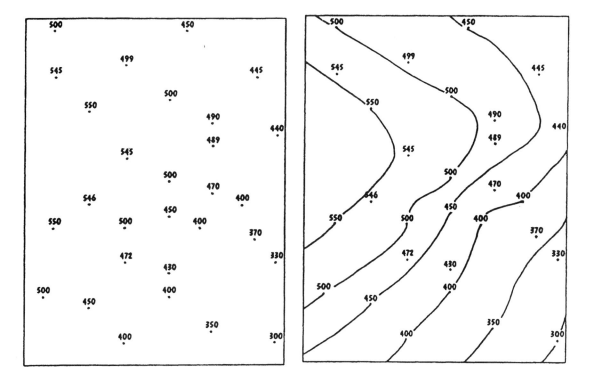

Figure 3 Known points of elevation are shown at the left. Contour lines from known points of elevation are shown at the right. The contour interval = 50 feet.

The vertical difference in elevation between contour lines is called the **contour interval**. The contour interval is usually stated at the lower margin of a topographic map. If it is not stated, look for a numbered contour line—an *index contour line* that is thicker and darker. Every fifth line is usually an index line. The choice of contour interval depends on the scale of the map.

All contour lines are multiples of the contour interval. For a contour interval of 10 feet the sequence of contours would be 0, 10, 20, 30, 40, etc. For a contour interval of 50 feet the sequence of contours would be 0, 50, 100, 150, 200, etc. If a point lies on a numbered contour line, the elevation is simply read from the line. If the contour line is unnumbered, the elevation can be found by counting the number of lines (contour intervals) above or below the index contour. Add or subtract this number from the index contour. If a point lies between contour lines, the elevation is approximated from the two contour lines it lies between. Elevations of specific points, such as mountain peaks, road crossings, or survey stations can sometimes be read directly from the map. Most of these points are marked with a small "x" and a number for the elevation.

RULES FOR CONTOUR LINES
1. All points on a contour line must be of equal elevation.
2. Contour lines always close to form a closed path, like an irregular circle. If the contour line extends beyond the mapped area the entire circle will not be seen.
3. Contour lines never cross one another.
4. Contour lines must never divide or split.
5. Closely spaced contour lines represent a steep slope; widely spaced contour lines represent a gentle slope.
6. A concentric series of closed contours represents a hill.
7. A concentric series of closed contours with hachure marks directed inward represent a closed depression.
8. Contour lines form a V pattern when crossing streams. The apex of the V always points upstream (uphill).

Summing Up Exercises

1. Draw in all necessary contour lines below. Use a contour interval of 10 meters.

135	122	110	88	100	105	90
140	125	113	94	120	121	100
143	132	120	108	127	135	140
150	135	128	120	140	144	155
156	138	134	138	147	156	165
160	146	141	153	155	165	180

2. Draw in all necessary contour lines below. Use a contour interval of 25 meters.

165	124	75	63	123	187	198
132	100	99	67	71	114	148
116	148	169	102	60	63	67
115	176	197	142	100	69	54
101	147	206	187	149	113	70
93	123	200	203	148	87	159
82	98	154	205	185	193	175
60	101	166	169	156	153	199
56	95	164	132	141	160	173

3. Mark each contour line with its corresponding elevation. Use a contour interval of 40 feet.

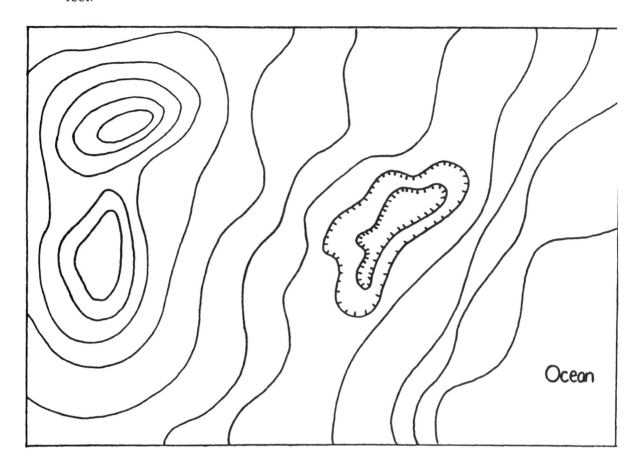

Topographic Profile

A topographic map provides an overhead view of the landscape. A profile of the landscape across any area of the map can be constructed by drawing a straight line across the desired area. This profile is like a slice through the landscape allowing us a side view. Figure 4 shows a method for constructing a topographic profile.

Procedure

Step 1: Mark a straight line across the map to indicate the line of profile; label it A-A'. Placement of the line can be anywhere, depending on what part of the landscape you want to view.

Step 2: Lay the edge of a sheet of paper along the line A-A'. Mark the position of each contour on the paper. Note the elevation for each mark. Other features such as mountain crests or the location of streams should also be marked.

Step 3: Prepare a graph with the line A-A' as the horizontal axis and elevation as the vertical axis. Each line will represent a contour line. Each line on the graph must be equally spaced. Label the lines so that the highest and lowest elevations along the line of profile fit into the graph. The vertical axis can be made to the same scale as the map or, for a more impressive profile, the scale can be exaggerated. Note: The size of exaggeration must be stated for correct interpretation of the profile.

For vertical exaggeration, first decide on a vertical scale.

For example: 1 inch = 100 feet = 1200 inches (1:1200). To calculate the vertical exaggeration, divide the fractional vertical scale by the fractional horizontal scale.

Example: For a vertical scale of 1:10,000 and a horizontal scale of 1:50,000 the vertical exaggeration will be 5 times greater than the true relief.

Step 4: Draw the profile on the graph by transferring the marked position of each contour elevation on its respective line. Connect the points with a smooth line.

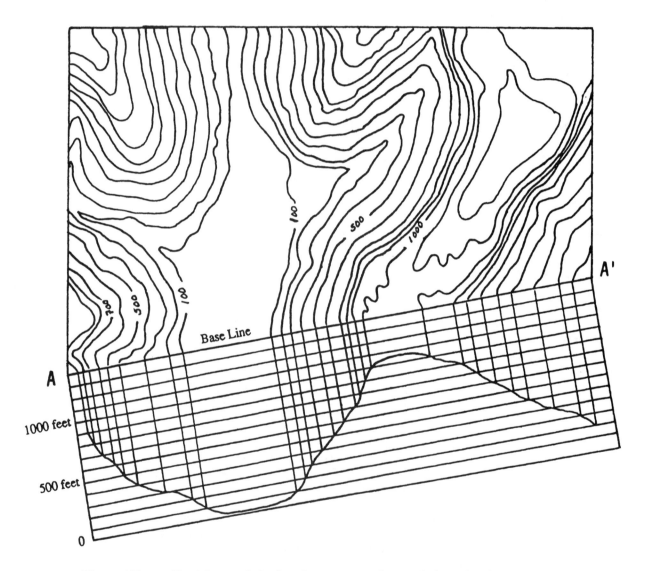

Figure 4 The profile of the terrain is plotted to an appropriate scale from the chosen baseline.

4. Construct a topographic profile for line A-A' below. Calculate the vertical exaggeration.

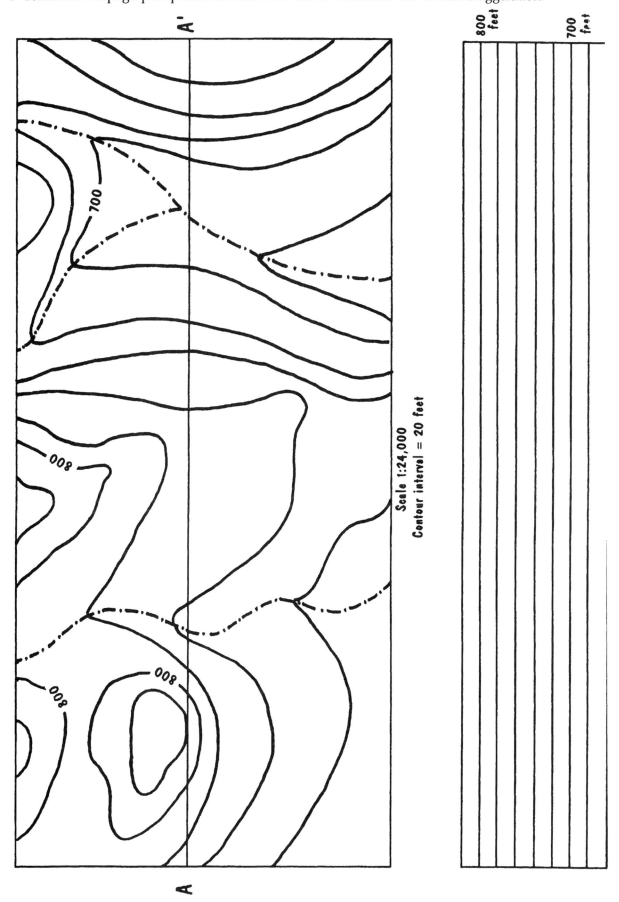

Scale 1:24,000
Contour interval = 20 feet

800 feet

700 feet

Name_____ Period_____ Date_____

Architecture of the Earth: Geologic Cross Sections

Over and Under

Purpose
In this activity, you will construct and interpret maps of geologic cross sections in the subsurface.

Required Equipment and Supplies
protractor
compass
colored pencils and eraser

Discussion
Geologic cross sections show the three-dimensional structure of the subsurface. They provide a "pie slice" of the subsurface and are thus extremely useful for mineral, ore, and oil exploration. We can construct geologic cross sections by using surface information—outcrops of folded and faulted rock sequences, and igneous rock intrusions.

The study of rock deformation is called *structural geology*. When a rock is subjected to compressive stress it begins to buckle and fold. If the stress overcomes the strength of the rock, the rock exhibits strain and breaks or faults. Structural geology interprets the different types of stress and strain.

We measure the orientation of deformed rock layers using strike and dip (Figure 1). **Strike** is the trend or direction of a horizontal line in an inclined plane. The direction of strike is expressed relative to north. For example, "North X degrees west," or "North X degrees east." **Dip** is the vertical angle between the horizontal plane and an inclined plane. Dip is always measured perpendicular to strike. We can think of the direction of dip as the direction a marble would roll down a plane. Hence, we express dip as (1) the direction a marble would roll, and (2) the angle between the inclined and horizontal planes. In Figure 1 the dip direction and angle are expressed as "45° west".

Fig. 1 Strike and dip. On the rock outcrop, the strike is the line formed from the intersection of a horizontal plane and the tilted rock strata. The dip is the angle between the horizontal and tilted strata (plane). The direction of dip is simply the geographical (N,S,E, or W) direction in which a marble would roll down the tilted plane. Dip is always measured perpendicular to strike. In the example shown, the outcrop is striking *northwest* and dipping *45° southwest*.

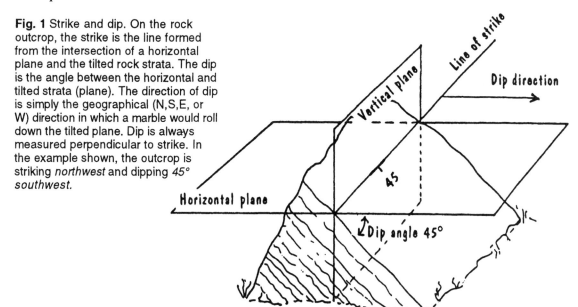

141

Geologic map symbols and symbols for strike and dip are shown in Figure 2.

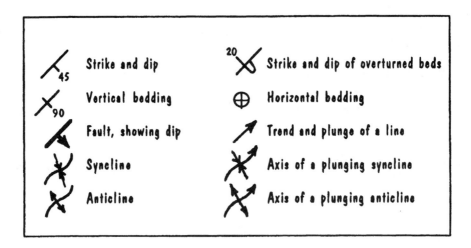

Fig. 2 Geologic map symbols.

The orientation of deformed layers can also be obtained through age relationships. Recall that sediments that settle out of water, such as in an ocean or in a bay, are deposited in horizontal layers. The layer at the bottom was deposited first and is therefore the oldest in the sequence of layers. Each new layer is deposited on top of the previous layer. Therefore in a sequence of sedimentary layers the oldest layer is at the bottom of the sequence and the youngest layer is at the top of the sequence. As these sedimentary layers are subjected to stress they fold and tilt. Each fold has an axis. If the tilted layers dip toward the fold axis, the fold is called a **syncline**. The rocks in the center, or core, are younger than those away from the core. If the tilted layers dip away from the axis, the fold is called an **anticline**. The rocks in the core of the fold of an anticline are older than the rocks away from the core (Figure 3).

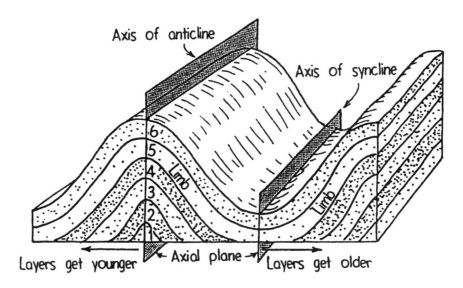

Fig. 3 Anticline and syncline folds.

The fold axis itself can be folded, but more often it is simply tilted. Folds in which the axis is tilted are known as *plunging folds* (Figure 4).

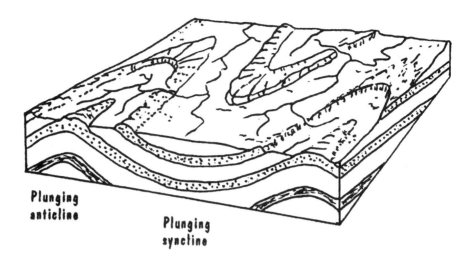

Fig. 4 Plunging folds.

When stress exceeds the mechanical strength of a rock, the rock breaks or faults. The type of fault can be deciphered from the ages of the rocks on either side of the fault. Thrust and reverse faults are produced by compression (squeezing) forces. These types of faults have older, structurally deeper rocks pushed on top of younger, structurally higher rocks. Normal faults are produced by tensional (pulling) forces. This type of fault has younger, structurally higher rocks on top of older, structurally deeper rocks. The field evidence for normal faults is repeated layers. The evidence for faults at the earth's surface is "missing" or "repeated" layers of rock.

Shearing forces produce strike-slip faults where the rocks slide past one another. Strike slip faults show no vertical movement, the movement is horizontal. The different types of faults are illustrated in Figure 5.

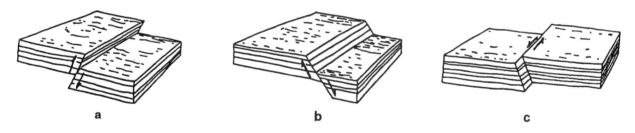

Fig. 5
a Compressive forces produce thrust faults and reverse faults;
b Tensional forces produce normal faults;
c Shearing forces produce strike-slip faults.

Age sequence relationships can also be determined from faults and igneous intrusions. When igneous intrusions or faults cut through other rocks, the intrusion or fault must be younger than the rocks they cut.

In this exercise we will use age relationships, strike and dip orientations, and fault evidence to construct cross sections from map views. A map view only shows the surface features—rock type, sequence of rocks, strike and dip, and fault lines. The cross section view is the subsurface extension of the surface view.

143

Procedure
Part A. The Construction of a Cross Section using Strike and Dip Measurements

Step 1: Refer to Figure 6. Transfer locations of contacts and strike and dip symbols to the topographic profile.

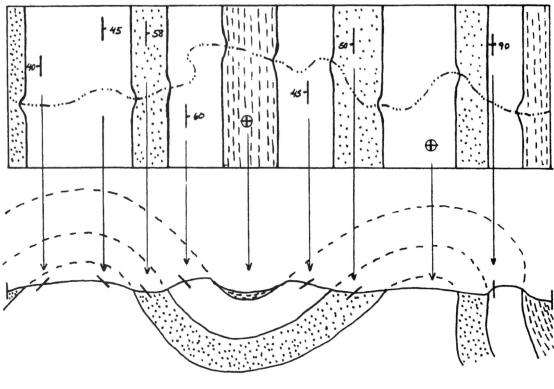

Fig. 6

Step 2: Align the protractor to the location of the transferred dip mark on the topographic profile. Depending on the dip direction, rotate the protractor to the measured angle. Mark this angle onto profile. Repeat for all dip marks.

Step 3: Using the dip lines on the topographic profile as guides, draw in the bedding contacts. The lines should be smooth and parallel.

Step 4: Dashed lines may be used to show the eroded surface.

Part B. The Construction of Cross Sections using Age Relationships

Step 1: Determine if there is evidence of folding. Folding will show a mirror image about the fold axis of geologic layers.

Step 2: If the beds are folded, determine if the bedding gets younger or older away from the axis of the fold.

Step 3: If the beds are not folded the disturbed sequence may be due to faulting. Are there repeated layers? Are there missing rock layers? Are the layers offset?

Part C. The Construction of Cross Sections Disturbed by Igneous Intrusions

Step 1: Refer to Figure 7. If the beds are not folded, look to see if there has been a disturbance—faulting or igneous intrusion. Faults and intrusions are always younger than the rock they cut into.

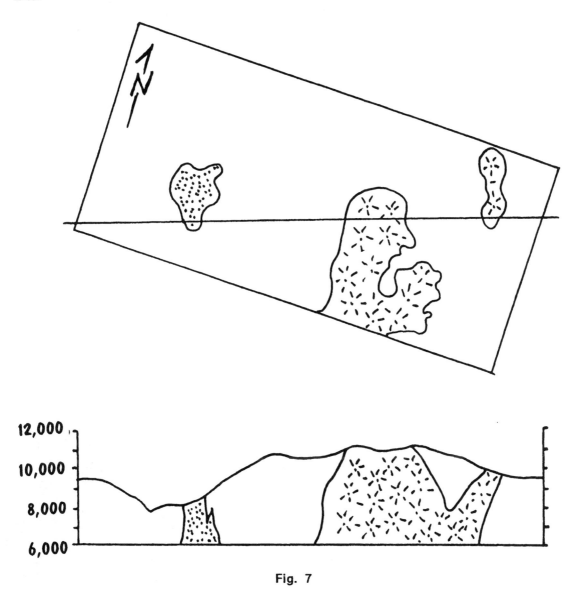

Fig. 7

Step 2: Igneous intrusions work their way up from the deep subsurface. Since there is no way to tell the extent of an intrusion in the subsurface, there is a lot of interpretation.

Exercises

1. Use strike and dip symbols to determine the cross section. What is the structure?

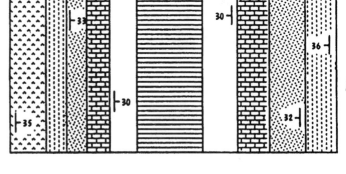

2. Use strike and dip symbols to determine the cross section. What type of structure is this?

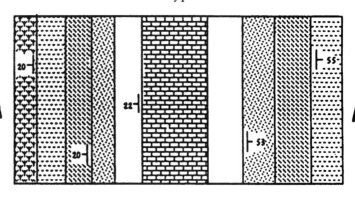

3. Use age relationships and general strike and dip symbols to determine the cross section. What type of structure is this?

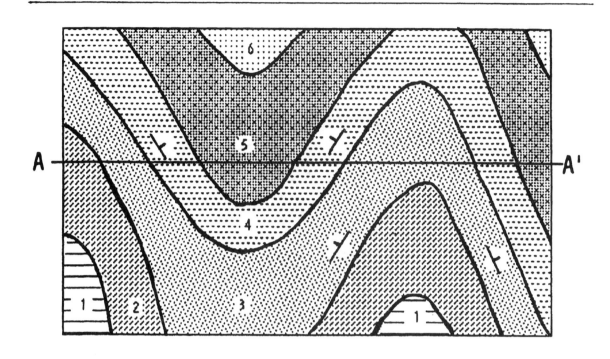

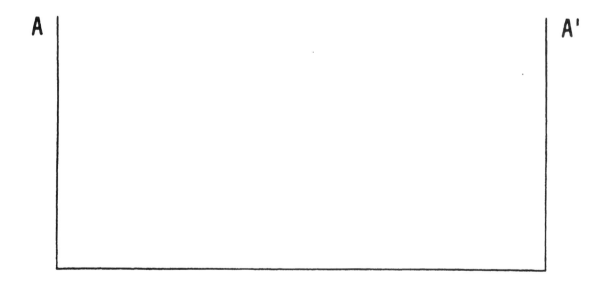

4. Use age relationships to determine the cross section. Note that 1 is the oldest layer and 6 is the youngest. What type of structure is this?

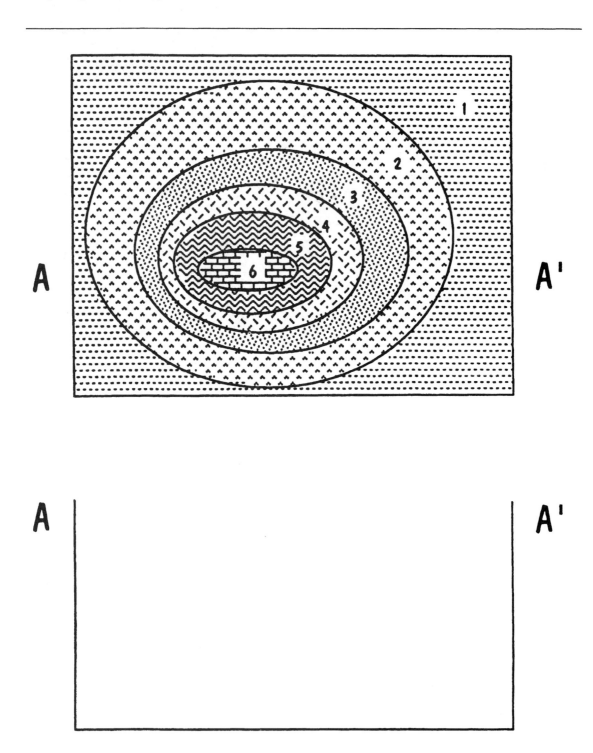

5. Use age relationships to determine the cross section. Note that 1 is the oldest layer and 5 is youngest. What type of structure is this?

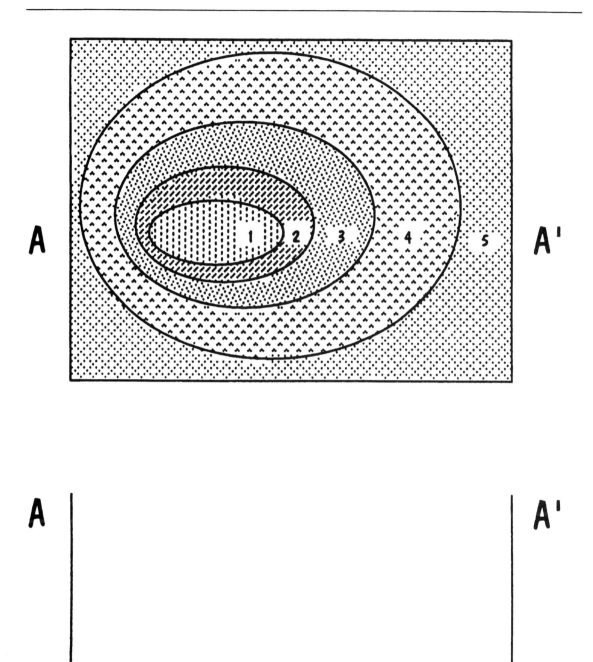

6. This map shows evidence of disturbance and will require some interpretation. Determine the fold structure using the general strike and dip symbols. What type of fold is displayed?

_____

7. The diagonal line represents a fault surface. Show the fault direction movement by placing arrows on the map. What type of fault does this represent?

_____

8. Draw the cross section. What is the oldest structure?

_____

9. What is the youngest structure?

_____

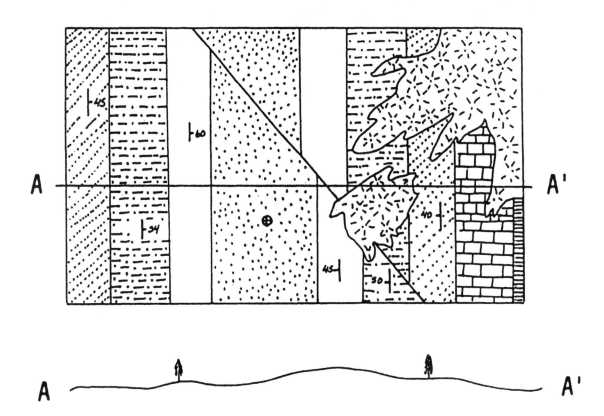

Summing Up

1. What type of structure is displayed in Exercise 1? Is the structure symmetrical or asymmetrical? Can the symmetry be determined from the map view?

_____

_____

_____

2. What type of structure is displayed in Exercise 2? Is the structure symmetrical or asymmetrical? Can the symmetry be determined from the map view?

_____

_____

_____

3. What type of structure is displayed in Exercise 3? What does the dip direction tell us about the structure?

_____

_____

_____

4. Exercise 4 and Exercise 5 are very similar to each other. How are they similar? How are they different? Mark the dip direction for both exercises.

_____

_____

_____

5. What type of fault is displayed in Exercise 6? What evidence supports your answer? What is the fold structure? Based on the drawing, which structure is oldest? Which is youngest?

_____

_____

_____

Water on Our World: Contour Maps of the Water Table

Walking on Water

Purpose

In this activity, you will construct contour maps of the water table and determine flow paths. Data for map construction can be water table levels obtained from wells and other techniques. You will construct contour lines from data. Your contour lines represent the water table in much the same way as contour lines represent land surface.

Required Equipment and Supplies

ruler
pencil and eraser
protractor

Discussion

Below the surface of continents is an extensive and accessible reservoir of fresh water. This reservoir of subsurface water is divided into two classes—soil moisture and groundwater. Groundwater is water that has percolated into the subsurface and saturated the open pore spaces in the rock or soil. The upper boundary of this saturated zone is called the **water table**. Water in the unsaturated zone above the water table is called *soil moisture*. The depth of the water table varies with precipitation and climate.

The water table tends to be a subdued version of the surface topography—sort of an underground surface. Water table contour maps, similar to surface contour maps, show the direction and speed of groundwater flow—information extremely useful for water supply management. For instance, simulated water table elevations generated by computer models can be compared with actual water table elevations in order to calibrate and verify the model. A groundwater model will tell you the best location for a well, and the impact that pumping from a new well will have on current water levels. Of particular interest today is the use of groundwater modeling to monitor contaminant transport in the subsurface.

Groundwater flow is influenced by gravitational force. Groundwater flows "downhill" underground, but the path it takes is dependent on *hydraulic head* and not topography. Hydraulic head is higher where the water table is high, such as beneath a hill, and lower where the water table is low, such as beneath a stream valley. So, responding to the force of gravity, water moves from high areas of the water table to low areas of the water table. As groundwater flows, it takes the path of least resistance—the shortest route. For any two lines of equal elevation, the shortest distance between them will be perpendicular to the lines. So, if we think of the path of groundwater flow as a *flow line*, the flow line will always be perpendicular to the contour lines.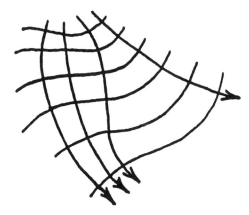

Problem 1: Construction of Water Table Contour Lines

The construction of contour lines for a water table is very similar to the construction of surface topographic contour lines. The same procedures and rules for surface contours apply to water table contour maps. The only difference is that you are drawing contour lines of the water table below the ground's surface.

Draw contour lines for the elevation of the water table. Use a contour interval = 10 feet.

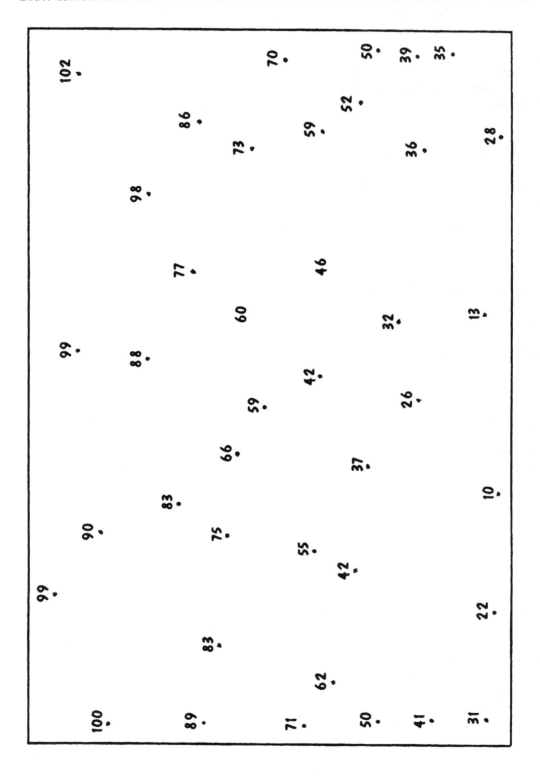

Problem 2: Construction of Flow Lines

Water flows from areas of high hydraulic head to areas of lower hydraulic head. In mapping groundwater flow, the key concept is that the line of flow is perpendicular to the contour lines of the water table. You are provided with the cross section view as a guide to your thinking.

Draw flow lines on the water table contour maps with arrows to show flow direction. Also draw arrows on the surface of the water table at cross sections to indicate the direction of flow.

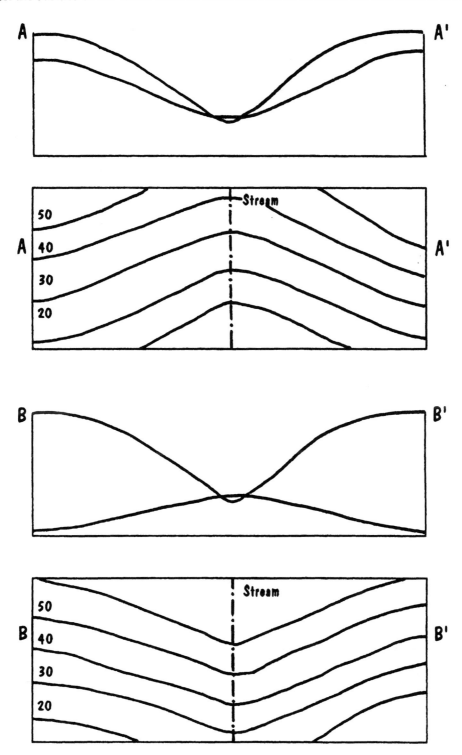

Knowledge of groundwater flow is important for solving problems of groundwater contamination. The most common groundwater contamination comes from sewage—drainage from septic tanks, inadequate or leaking sewer pipes, and farm waste areas. Contamination may also result from landfills and waste dumps where toxic materials and hazardous wastes leach down into the subsurface.

Problem 3: Contaminant Flow Model
Draw contour lines for the elevation of the water table. Use a contour interval of 10 feet. Water wells and a sewer treatment facility have been plotted on the map. The treatment facility is in disrepair and has developed a leak. Determine which well (if any) will be affected by the flow of contamination from the sewage treatment facility.

Summing Up

1. In Problem 2, what is the difference between stream A and stream B? In what type of climatic area (dry desert or wet forest) would we likely find stream A? In what type of climatic area would we likely find stream B? Explain.

_____

_____

_____

2. In Problem 3 we are assuming equal rates of pumping for each well. If excessive pumping occurred at well H, would there be a change in the flow of contamination? Which well, if any, would likely become contaminated because of this pumping?

_____

_____

_____

3. In Problem 3 we are assuming that all the soil in the subsurface is homogeneous (the same). If an impermeable clay lens were found at point 55, would the flow of sewage contamination be affected? How about point 35?

_____

_____

_____

| CONCEPTUAL PHYSICAL SCIENCE **EXPLORATIONS** | **Experiment** |

Atmosphere and Ocean Interactions: Solar Energy

Solar Power I

Purpose
In this experiment, you will measure the sun's power output by comparison with the power output of a 100-watt light bulb.

Required Equipment and Supplies
2-cm by 6-cm piece of aluminum foil with one side thinly coated with flat black paint
clear tape
meterstick
glass jar with a hole in the metal lid
one-hole stopper to fit the hole in the jar lid
fine-scaled thermometer
clear 100-watt light bulb with receptacle

Discussion
To measure the power output of a light bulb is nothing to write home about—but to measure the power output of the sun using only household equipment is a different story. In this experiment we'll do just that—measure to a fair approximation the power output of the sun. We'll do this by comparing the light from a bulb of known wattage with the light from the sun, using the simple ratio

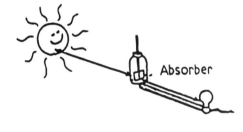

Absorber

$$\frac{\text{Sun's wattage}}{\text{Sun's distance}^2} = \frac{\text{bulb's wattage}}{\text{bulb's distance}^2}$$

We'll need a sunny day to do this experiment.

Procedure
Step 1: With the blackened side facing out, fold the middle of the foil strip around the thermometer bulb, as shown in Figure A. The ends of the metal strip should line up evenly.

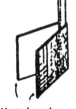

Metal ends even
Fig. A

Step 2: Crimp the foil so it completely surrounds the bulb, as in Figure B. Bend each end of the foil strip outward (Figure C). On a tabletop or with a meter stick, make a flat, even surface. Use a piece of clear tape to hold the foil to the thermometer.

Crimp here
Fig. B

Step 3: Insert the free end of the thermometer into the one hole-stopper (soapy water or glycerin helps). Remove the lid from the jar, and place the stopper in the lid from the bottom side. Slide the thermometer until the foil strip is located in the middle of the jar. Place the lid on the jar.

Step 4: Position the jar indoors near a window so that the sun will shine on it. Prop it at an angle so that the blackened side of the foil strip is perpendicular to the rays of the sun. Keep it in this position in the sunlight until a stable temperature is reached. If you prefer, do this outside, which means doing Steps 5 and 6 outside.

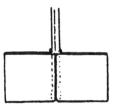

Bend blades outward
Fig. C

Stable temperature = _____ °C

Fig. D

Step 5: Now find the conditions for bringing the foil to the same temperature with a clear 100-watt light bulb. Set the meter-stick on the table. Place the clear 100-watt light bulb with its filament located at the 0-cm mark of the meter-stick (Figure D). Center the jar at the 95-cm mark with the blackened side of the foil strip perpendicular to the light rays from the bulb. You may need to put some books under the jar to align it properly.

Step 6: Turn the light bulb on. Slowly move the jar toward the light bulb, 5 cm at a time, allowing the thermometer temperature to stabilize each time. As the temperature approaches the reading reached in Step 4, move the jar only 1 cm at a time. When the bulb maintains the same temperature obtained from the sun for about two minutes, turn the light bulb off.

Step 7: Measure as exactly as possible the distance in meters between the foil and the filament of the bulb. Record this distance.

Distance from light filament to foil strip = _____ m

Step 8: Since you know the sun's distance from the thermometer in meters is 1.5×10^{11} m, and you know the distance of the light bulb from the thermometer, and the wattage of the bulb, you know 3 of the 4 values for the ratio equality stated earlier. With simply rearrangement

$$\text{Sun's wattage} = \frac{(\text{bulb's wattage})(\text{Sun's distance})^2}{(\text{bulb's distance})^2}$$

Show your work:

Sun's wattage = _____ W

Step 9: Use the sun's wattage to compute the number of 100-watt light bulbs needed to equal the sun's power. Show your work.

Number of 100-watt light bulbs = _____

Summing Up

The accepted value for the sun's power is 3.8×10^{26} W. How close was your experimental value? List factors you can think of that affect the difference.

_____

_____

_____

Name_____ Period_____ Date_____

Atmosphere and Ocean Interactions: Solar Energy

Solar Power II

Purpose
In this experiment, you will measure the amount of solar energy per minute that reaches the earth's surface, and from this estimate the sun's power output.

Required Equipment and Supplies
2 Styrofoam cups
graduated cylinder
water
blue and green food coloring
plastic wrap
rubber band
thermometer (Celsius)
meter stick

Discussion
How do we know how much energy the sun radiates? First we assume that the sun radiates energy equally in all directions. Imagine a heat detector so big that it completely surrounds the sun, like an enormous basketball with the sun at its center. Then the heat energy reaching this detector each second would be the sun's power output. Or if our detector were half a basketball and caught half the sun's energy, then we would multiply the detector reading by two to compute the total solar output. If our detector were a quarter of a basketball, and caught one quarter of the sun's energy, then its total output would be four times the detector reading.

Now that you have the concept, suppose that our detector is the water surface area of a full Styrofoam cup here on earth facing the sun. Then that area is only a tiny fraction of the area surrounding the sun. If we figure what that fraction is, and also measure the energy captured by the water in our cup, we can tell how much total energy the sun radiates. In this experiment, we'll measure the amount of solar energy that reaches a Styrofoam cup and relate it to the amount of solar energy that falls on the earth. We'll need a sunny day for this.

Procedure
Step 1: Using a graduated cylinder, measure and record the amount of water needed to fill a Styrofoam cup. Add a small amount of blue and green food coloring to make the water dark (and a better absorber of solar energy).

Volume of water = _____ mL

Mass of water = _____ g

Nest the water-filled cup in a second Styrofoam cup
(for better insulation).

Step 2: Measure the water temperature and record it.

Initial water temperature = _____ °C

Step 3: Cover the doubled cup with plastic wrap, and seal it with a rubber band to prevent water leaks. Place the cup in the sunlight, tipped if necessary so that the plastic surface is approximately perpendicular to the sun's rays. Let it warm by sunlight for 10 minutes.

Step 4: Remove the plastic wrap. Stir the water in the cup gently with the thermometer. Measure the final water temperature and record it. Find the difference in temperature before and after being in the sunlight.

THE QUANTITY Q OF HEAT ENERGY COLLECTED BY THE WATER = MASS OF THE WATER × ITS SPECIFIC HEAT (c= 1 $\frac{cal}{g°c}$) × ΔT, ITS CHANGE IN TEMPERATURE

Final water temperature = _____ °C

Temperature difference = _____ °C

Step 5: Compute the surface area of the top of the cup in square centimeters. (Begin by measuring and recording the diameter of the cup's top in centimeters).

Cup diameter = _____ cm

Surface area of water = _____ cm^2

Step 6: Compute the energy in calories that was collected in the cup ($Q = mc\Delta T$). Assume that the specific heat of the mixture is the same as the specific heat of the water. Show your work on separate paper and record your result here.

Energy = _____ cal

Step 7: Compute the solar energy flux, the energy collected per square centimeter per minute.

Solar energy flux = _____ $\dfrac{cal}{cm^2min}$

Step 8: Compute how much solar energy reaches each square meter of the earth per minute. Again, show your work on separate paper. (Hint: There are 10,000 cm^2 in 1 m^2.)

Solar energy flux = _____ $\dfrac{cal}{m^2min}$

Step 10: The distance between the earth and the sun is 1.5 x10^{11}m, or 1.5 x10^{13}cm. The area of a sphere is $4\pi r^2$ so you can calculate the surface area of a "basketball" with radius 1.5 x 10^{13}cm. And we can divide this area by the surface area of the cup we used in this experiment to find what fraction of the sun's total solar output reached this cup. Then you can calculate the total solar output per minute from your work in Step 7. Do so.

Total solar output per minute = _____ cal

Summing Up

1. Scientists have measured the amount of solar energy flux just above our atmosphere to be 2 calories per square centimeter per minute (equivalently 1.4 kW/m^2). This energy flux is called the *solar constant*. Only 1.5 calories per square centimeter per minute reaches the earth's surface after passing through the atmosphere. How did your value compare?

_____

2. What factors could affect the amount of sunlight reaching the earth's surface, decreasing the solar constant?

_____

CONCEPTUAL PHYSICAL SCIENCE **EXPLORATIONS** | **Activity**

Weather: Clouds

Indoor Clouds

Purpose
In this activity, you will illustrate the formation of a cloud from the condensation of water droplets.

Required Equipment and Supplies
large glass jar
measuring cup for water
small metal baking tray
tray of ice cubes

Discussion
Clouds are made up of millions of tiny water droplets and/or ice crystals. Cloud formation takes place as rising moist air expands and cools.

As the sun warms the earth's surface, water is evaporated from the oceans, lakes, streams, and rivers. The process of evaporation changes the water molecules from a liquid phase to a vapor phase. Because evaporation is greater over warmer waters than cooler waters, tropical locations have a higher water vapor content than polar locations. Water vapor content is dependent on temperature. For any given temperature there is a limit to the amount of water vapor in the air. This limit is called the dew point. When this limit is reached the air is saturated. A measure of the amount of water vapor in the air is called humidity (the mass of water per volume of air). The amount of water vapor in the air varies with geographic location and may change according to temperature.

Warm air can hold more water vapor before becoming saturated than cooler air. When the air temperature falls, the air becomes saturated—the amount of water vapor the air can hold reaches its limit. As the air cools below the dew point, the water vapor molecules condense onto the nearest available surface. Condensation is the change from a vapor phase to a liquid phase. The condensation of water vapor molecules on small airborne particles produces cloud droplets, which in turn become clouds.

Most clouds form as air rises, expands, and cools. There are several reasons for the development of clouds. When the temperatures of certain areas of the earth's surface increase more readily than other areas, air may rise from thermal convection. As this warm air rises, it mixes with the cooler air above and eventually cools to its saturation point. The moisture from the warm air condenses to form a cloud. As the cloud grows, it shades the ground from the sun. This cuts off the surface heating and upward thermal convection—the cloud dissipates. After the cloud is gone, the ground once again heats up to start another cycle of thermal convection.

Clouds also form as a result of topography. Air rises as it moves over mountains. As it rises, it cools. If the air is humid, clouds form. As the air moves down the other side of the mountain, it warms. This air is drier since most of the moisture has been removed to form the clouds on the other side. It is therefore more common to find cloud formation on the windward side of mountains rather than on the leeward side.

Clouds may form as a result of converging air. When cold air moves into a warm air mass, the warm air is forced upward. As it rises, it cools, and water vapor condenses to form clouds. Cold fronts are associated with extensive cloudiness and thunderstorms. When warm air moves into a cold air mass, the less dense warmer air rides up and over the colder, denser air. Warm fronts result in widespread cloudiness and light precipitation that may extend for thousands of square kilometers.

Procedure

Step 1: Fill a glass jar with about one inch of very hot water. Do not use boiling water—the glass may break.

Step 2: Place ice cubes in a metal baking tray and set the tray on top of the jar. Make sure there is a good seal.

Step 3: Observe a "cloud." As the ice above cools the air inside the top of the jar, water vapor condenses into water droplets to form a "cloud."

Summing Up

1. How is your formation of clouds the same as the formation of clouds in the sky?

_____

_____

2. How is it different?

_____

_____

3. Why are coastal tropical areas more humid than desert areas?

_____

_____

4. Why do we believe that warm air rises?

_____

_____

5. How does topography contribute to the formation of desert areas? For example, what role do the Sierra Nevada Mountains have in keeping the Nevada Desert.

_____

_____

CONCEPTUAL PHYSICAL SCIENCE EXPLORATIONS | **Experiment**

The Solar System: Size of the Sun

Sunballs

Purpose
In this experiment, you will estimate the diameter of the sun.

Required Equipment and Supplies
small piece of cardboard
meterstick

Discussion
Take notice of the round spots of light on the shady ground beneath trees. These are sunballs—images of the sun (discussed in the Prologue of *Conceptual Physical Science— Explorations*). They are cast by openings between leaves in the trees that act as pinholes. The diameter of a sunball depends on its distance from the small opening that produces it. Large sunballs, several centimeters or so in diameter, are cast by openings that are relatively high above the ground, while small ones are produced by closer "pinholes." The interesting point is that the ratio of the diameter of the sunball to its distance from the pinhole is the same as the ratio of the sun's diameter to its distance from the pinhole.

Since the sun is approximately 150,000,000 km from the pinhole, careful measurement of this ratio tells us the diameter of the sun. That's what this experiment is all about. Instead of finding sunballs under the canopy of trees, you'll make your own easier-to-measure sunballs.

Procedure
Poke a small hole in a piece of cardboard with a pen or sharp pencil. Hold the cardboard in the sunlight and note the circular image that is cast on a convenient screen of any kind. This is an image of the sun. Unless you're holding the card too close, note that the solar image size does not depend on the size of the hole in the cardboard (pinhole), but only on its distance from the pinhole to the screen. The greater the distance between the image and the cardboard, the larger the sunball.

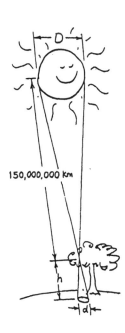

150,000,000 km

Position the cardboard so the image exactly covers a dime, or something that can be accurately measured. Carefully measure the distance to the small hole in the cardboard. Record your measurements as a ratio:

$$\frac{\text{diameter of dime}}{\text{distance from dime to pinhole}} = \underline{\hspace{3cm}}$$

Since this is the same ratio as the diameter of the sun to its distance, then

$$\frac{\text{diameter of dime}}{\text{distance from dime to pinhole}} = \frac{\text{diameter of sun}}{\text{distance from sun to pinhole}}$$

Which means you can now calculate the diameter of the sun!

Diameter of the sun = _____

165

Summing Up

1. Will the sunball still be round if the pinhole is square shaped? Triangular shaped? (Experiment and see!)

_____

_____

_____

2. If the sun is low so the sunball is elliptical, should you measure the small or the long width of the ellipse for the sunball diameter in your calculation of the sun's diameter? Why?

_____

_____

_____

3. If the sun is partially eclipsed, what will be the shape of the sunball?

_____

_____

_____

WHAT SHAPE DO SUNBALLS HAVE DURING A PARTIAL ECLIPSE OF THE SUN?

CONCEPTUAL PHYSICAL SCIENCE EXPLORATIONS | **Activity**

The Solar System: Planetary Motion

Ellipses

Purpose
In this activity, you will investigate the geometry of the ellipse.

Required Equipment and Supplies
about 20 cm of string
2 thumbtacks
pencil and paper
a flat surface that will accept thumbtack punctures

Discussion
An ellipse is an oval-like closed curve defined as the locus of all points about a pair of foci (focal points), the sum of whose distances from both foci is a constant. Planets orbit the sun in elliptical paths, with the sun's center at one focus. The other focus is a point in space, typified by nothing in particular.

Elliptical trajectories are not confined to "outer space." Toss a ball and that parabolic path it seems to trace is actually a small segment of an ellipse. If extended, its path would continue through the earth and swing about the earth's center and return to its starting point. In this case, the far focus is the earth's center. The near focus is not typified by anything in particular. The ellipse is very stretched out, with its long axis considerably longer than its short axis. We

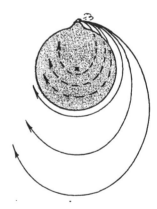

say the ellipse is very *eccentric*. Toss the rock faster and the ellipse is wider—less eccentric. The far focus is still the earth's center, and the near focus is nearer to the earth's center than before. Toss the rock at 8 km/s and both foci will coincide at the earth's center. The elliptical path is now a circle—a special case of an ellipse. Toss the rock faster, and it follows an ellipse external to the earth. Now we see the near focus is the center of the earth and the *far* focus is beyond—again, at no particular place. As speed increases and the ellipse becomes more eccentric again, the far focus is *outside* the earth's interior.

Constructing an ellipse with pencil, paper, string, and tacks, is interesting. Let's do it!

Procedure
Place a loop of string around two tacks or push pins and pull the string taut with a pencil. Then slide the pencil along the string, keeping it taut (Figure 2). (To avoid twisting of the string, it helps if you make the top and bottom half in two separate operations.)

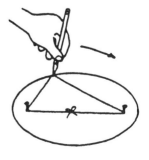

Label each focus of your ellipse (the location of the push pins). Repeat, using different focus separation distances, and you'll notice that the greater the distance between foci, the more eccentric will be the ellipse.

Construct a circle by bringing both foci together. A circle is a special case of an ellipse.

Another special case of an ellipse is a straight line. Determine the positions of the foci that give you a straight line.

The shadow of a ball face on is a circle. But when the shadow is not face on, the shape is an ellipse. The photo below shows a ball illuminated with three light sources. Interestingly enough, the ball meets the table at the focus for all three ellipses.

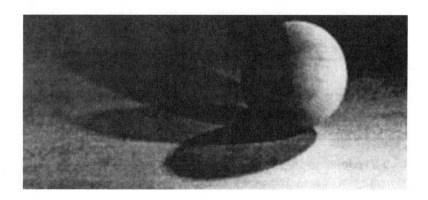

Summing Up

1. Which of your drawings approximates the earth's orbit about the sun?

_____

2. Which of your drawings approximates the orbit of Haley's comet about the sun?

_____

3. What is your evidence for the definition of the ellipse: that the sum of the distances from the foci is constant?

_____

_____

_____

Name_____ Period_____ Date_____

The Solar System: Astronomy Measurements

Reckoning Latitude

Purpose
In this experiment, you will build and use the necessary instruments for determining how far above the equator you are on this planet. This will be done by measuring the angle of Polaris in the sky. Actual measurements will need to be made outside of class since measurements can only be performed at nighttime.

Required Material and Supplies
protractor
pencil with unused eraser
pushpin
drinking straw
plumb line and bob
tape

Discussion
Latitude is a measure of your angular distance north or south of the equator (Figure 1). To determine your latitude in the northern hemisphere, it is convenient to use Polaris, the North Star, as a guide. At the North Pole you would find that the North Star is directly above you—90° above the horizon. This is also your angular distance from the equator (Figure 2a). If you were to travel closer to the equator, the North Star would recede behind you to some lower angle above the horizon, which would always match your angular distance from the equator (Figure 2b). If for example, you measured the North Star to be 45° above the horizon, you would be at an angular distance of 45° north of the equator. At the equator you'd find the North Star to be 0° above the horizon (Figure 2c). (It turns out atmospheric that refraction at this grazing angle complicates viewing.)

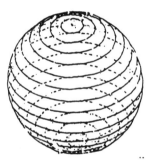

Fig. 1 Latitudes are parallel to the equator

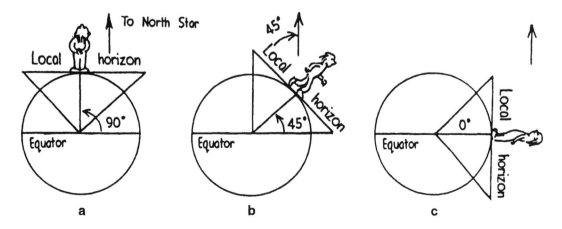

Fig. 2 The angle at which the North Star appears above the horizon is also your angular distance from the equator.

169

Procedure

An astrolabe is a simple instrument for measuring angles above the horizon (altitude). To construct an astrolabe, tape a straw to the straight edge of a protractor as shown in Figure 3. Place the eraser end of a pencil against the hole in the protractor and insert the pushpin through the other side of the hole into the eraser—the pencil serves as a convenient handle. Next, tie a plumb bob to the needle (a string attached to a weight).

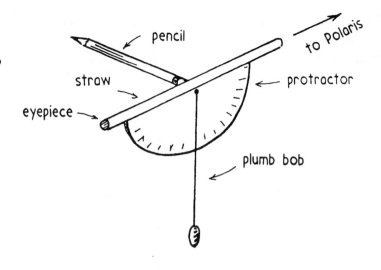

Fig. 3 Astrolabe for measuring altitude.

To measure the altitude of any object viewed through the straw, have a partner read the angle the plumb bob line makes with the protractor. Practice the use of your astrolabe by measuring the angle of various objects around you. Work in pairs or small groups so that one person can read the angle while the other is sighting. Take turns looking through the astrolabe to be sure of your readings. You may want to make several measurements and find the average. The angle that you read should be between 0° and 90°. Be sure that this angle is not the number of degrees from the zenith (straight up). You can convert such readings into degrees from the horizon (straight out) by subtracting the number of degrees from 90°. Challenge yourself or others to estimate measurements without using the astrolabe. How accurate are these "naked-eye" estimates?

Measuring the Altitude of Polaris

The North Star is fairly easy to locate because of its position relative to the Big and Little Dippers, two well-known constellations that stand out in the northern sky. To find the North Star from the Big Dipper, draw an imaginary line between the last two stars of the Big Dipper's pan. Follow this line away from the top of the Big Dipper and the first bright star you cross is the North Star (Figure 4). Note that the North Star is the first star of the Little Dipper's handle .

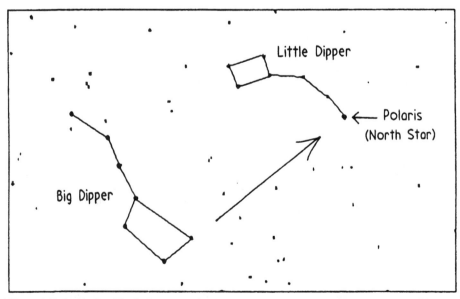

Fig. 4 Polaris, the North Star, can be found in the northern sky from its position relative to the Big and Little Dippers.

170

To find your latitude you need simply measure the altitude of the North Star above the horizon. It so happens that the North Star is not *exactly* above the North Celestial Pole, so for a more accurate determination of your latitude, a minor but significant correction may be necessary. The North Star lies about $3/4$ of a degree away from the North Celestial Pole in the direction of Cassiopeia, a "W"-shaped constellation (Figure 5). Hence the altitude of the North Star may not correspond exactly to your latitude. If you see Cassiopeia much "above" the North Star (higher up from the horizon), then subtract $3/4$ of a degree from your measured altitude of Polaris to obtain your latitude. If, on the other hand, Cassiopeia is "beneath" the North Star, add $3/4$ of a degree. If Cassiopeia is anywhere to the left or right of the North Star, then the correction is not necessary.

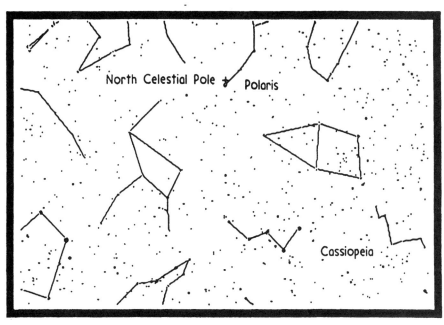

Fig. 5 Polaris is offset from the pole towards Cassiopeia.

Summing Up

1. What was your measured altitude of Polaris?

_____

2. According to this measurement, at what latitude are you located?

_____

3. By how many degrees does your measured latitude differ from the accepted latitude given by your instructor?

_____

4. One degree of latitude is equivalent to 107 km (66 miles). By how many kilometers does your determined latitude differ from the accepted value in kilometers?

   (Your Determined Latitude - Accepted Latitude) x 107 = _____

5. How could the astrolabe be used to provide evidence that the world is round?

_____

CONCEPTUAL PHYSICAL SCIENCE **EXPLORATIONS**	**Activity**

The Solar System: Planetary Motion

Tracking Mars

Purpose
In this activity, you will plot the orbit of Mars. You will employ the technique of Johannes Kepler and use data obtained by Tycho Brahe four centuries ago.

Required Equipment and Supplies
four sheets of plain grid graph paper
compass
protractor
ruler
sharp pencil

Discussion

In the early 1500s, the Polish astronomer Nicolaus Copernicus used observations and geometry to determine the orbital radius and orbital period for the planets known at the time. Copernicus based his work on the revolutionary assumption that the planets moved around the sun. In the late 1500s, before telescopes were invented, the Danish astronomer Tycho Brahe made twenty years of extensive and accurate measurements of planets and bright stars. Near the end of Tycho's career, he hired a young German mathematician, Johannes Kepler, who was assigned the task of plotting the orbit of Mars using Tycho's data.

Kepler started by drawing a circle to represent the earth's orbit (not bad, considering the earth's orbit only approximates a circle, which he didn't know at the time). Since Mars takes 687 Earth days to orbit the sun once, Kepler paid attention to observations that were exactly 687 days apart. In this way Mars would be in the same place while Earth would be in a different location. Two angular readings of Mars' location from the same location on Earth 687 days apart was all that was needed—where the two lines crossed was a point on the orbit of Mars. Plotting many such points did not trace out a circle, as Kepler had expected—rather, the path was an ellipse. Kepler was the first to discover that if the planets orbit the sun, they did so in elliptical rather than circular paths. He then went on to plot a better orbit for Earth based on observations of the sun, and further refined the plotted orbit of Mars.

In this activity you will duplicate the work of Kepler, simplifying somewhat by assuming a circular orbit for the earth—it turns out the difference is minor, and the elliptical path of Mars is evident. From Brahe's extensive tables, we'll use only data shown in Table A.

Procedure
Step 1: You'll want a sheet of graph paper with about a 14 x 14-inch working area. If need be, tape two legal-size sheets of borderless graph paper together, or four regular 8 1/2" x 11" sheets.

Step 2: Make a dot at the center of your paper to represent the sun. Place a compass there, and draw a 10-cm radius circle to represent the orbit of the earth around the sun. Draw a light line from the center to the right of the paper, and mark the intersection with the earth's orbit 0°. This is the position of the earth on March 21st. All your plotting will be counterclockwise around the circle from this reference point.

Table A Brahe's data are grouped in 14 pairs of Mars' sightings. For the first 9 pairs, the first line of data is for Mars in *opposition*—when Sun, Earth and Mars were on the same line—when Mars was 90° to the earth horizon, directly overhead at midnight. The second line of data are positions measured 687 days later, when Mars was again in the same place in its orbit, and Earth in a different place, where a different angle was then measured. All angles given in the table read from a 0° reference line—the line from the sun to Earth at the vernal equinox, March 21. Mars at points 10-14 are non-opposition sightings. The first line of Point 10, for example, shows that when Earth was at 277°, Mars was seen not directly overhead, but at 208.5° with respect to the 0° reference line. Then 687 days later Earth was at 235°, where Mars was seen at 272.5°. The data are neatly arranged for plotting—something that took Kepler years to do.

Mars Orbit Point	Date			Earth position (Ecliptic)	Mars Position (Ecliptic)
	Mo	Day	Year		
1	11	28	1580	66.5	66.5
	10	16	1582	22.5	107
2	1	7	1583	107	107
	11	24	1584	62.5	144
3	2	10	1585	141.5	141.5
	12	29	1586	97.5	177
4	3	16	1587	175.5	175.5
	1	31	1589	132	212
5	4	24	1589	214.5	214.5
	3	12	1591	171.5	253.5
6	6	18	1591	266.5	266.5
	5	5	1593	225	311
7	9	5	1593	342.5	342.5
	7	24	1595	300.5	29.5
8	11	10	1595	47.5	47.5
	9	27	1597	4	90
9	12	24	1597	92.5	92.5
	11	11	1599	48	130.5
10	6	29	1589	277	208.5
	5	16	1591	235	272.5
11	8	1	1591	308	260.5
	6	18	1593	266.5	335
12	9	9	1591	345.5	273
	7	27	1593	304	347.5
13	10	3	1593	9.5	337.5
	8	20	1595	327	44.5
14	11	23	1593	60.5	350.5
	10	10	1595	17	56

Step 3: Locate the first point in Mars' orbit, Point 1, from Table A. Do this by first marking with a protractor the position of the earth along the circle for the date November 28, 1580. This is 66.5° above the 0° reference line. Draw a dot to show Earth's position at this time.

Step 4: Mars at this time was in *opposition*—opposite to the sun in the sky. A line from the sun to Earth at this time extends radially outward to Mars. Draw a line from the center of your circle (the sun) to the earth's position at this time, and beyond the earth through Mars. Where is Mars along this line? You'll need another sighting of Mars 687 days later when Mars is at the same place and Earth is in another.

Step 5: If you were to add 687 days to November 28, 1580, you'd get October 16, 1582. At that date Mars was measured to be lower in the sky—actually 107.0° with respect to the 0° reference line of March 21. With respect to the reference line, use a protractor and ruler and draw a line

174

at 107.0° as shown in Figure A. Where your two lines intersect is a point along Mars' orbit.

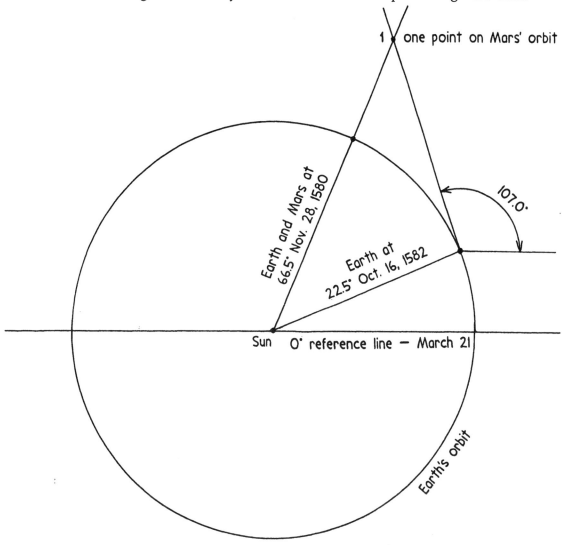

Fig. A Your plot should look like this for finding Mars Orbit Point 1; data are in the first pair of lines of Table A. Angles from Earth to Mars are given with respect to the 0° reference line of March 21.

Step 6: With care, plot the 13 other intersections that represent points along Mars' orbit, using the data in Table A.

Step 7: Connect your points, either very carefully by freehand, or with a French curve.

Bravo—you have plotted the orbit of Mars using Kepler's method from four centuries ago!

Summing Up

1. Place the point of your compass on the position of the sun and place the pencil on the first point of the plotted orbit of Mars. Draw a circular orbit with that radius. Does that circle match the orbit of Mars? If not, describe the differences.

_____

_____

2. Does your plot agree with Kepler's finding that the orbit is an ellipse? _____

3. During what month are the orbits of Mars and Earth closest? _____

175

Appendix A
Significant Figures and Uncertainty in Measurement

Units of Measurement
All measurements consist of a number and a unit. Both are necessary. If you say that a friend is going to give you 5, you are telling only *how many*. You also need to tell *what*—five fingers, five cents, five dollars, or five corny jokes. If your instructor asks you to measure the length of a piece of wood, saying that the answer is 26 is not correct. She or he needs to know whether the length is 26 centimeters, feet, or meters. All measurements must be expressed using a number and an appropriate unit.

Numbers
Two kinds of numbers are used in science—those that are counted or defined, and those that are measured. There is a great difference between a counted or defined number and a measured number. The exact value of a counted or defined number can be stated, but the exact value of a measured number cannot.

For example, you can count the number of chairs in your classroom, the number of fingers on your hand, or the number of quarters in your pocket with absolute certainty. Counted numbers are not subject to error (unless you counted wrong).

Defined numbers are about exact relations and are defined to be true. The defined number of centimeters in a meter, the defined number of seconds in an hour, and the defined number of sides on a square are examples. Defined numbers also are not subject to error (unless you forget the definition).

Every measured number, no matter how carefully measured, has some degree of uncertainty. What is the width of your desk? Is it 89.5 centimeters, 89.52 centimeters, 89.520 centimeters or 89.5201 centimeters? You cannot state its exact measurement with absolute certainty.

Uncertainty in Measurement
Uncertainty (or margin of error) in a measurement can be illustrated by the two different metersticks in Figure A. The measurements are of the length of a tabletop. Assuming that the zero end of the meterstick has been carefully and accurately positioned at the left end of the table, how long is the table?

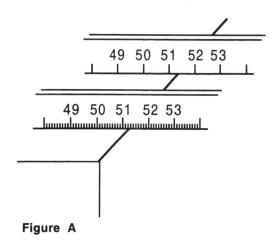

The upper scale in the figure is marked off in centimeter intervals. Using this scale you can say with certainty that the length is between 51 and 52 centimeters. You can say further that it is closer to 51 centimeters than to 52 centimeters; you can estimate it to be 51.2 centimeters.

Figure A

The lower scale has more subdivisions and has a greater precision because it is marked off in millimeters. With this meterstick you can say that the length is definitely between 51.2 and 51.3 centimeters, and you can estimate it to be 51.25 centimeters.

Note how both readings contain some digits that are known, and one digit (the last one) that is estimated. Note also that the uncertainty in the reading of the lower meterstick is less than that of the top meterstick. The lower meterstick can give a reading to the hundredths place, and the top meterstick to the tenths place. The lower meterstick is more *precise* than the top one. So, digits tell us the magnitude of a measurement while the location of the decimal point tells us the precision.

Significant Figures

Significant figures are the digits in any measurement that are known with certainty plus the one digit that is estimated and hence is uncertain. The measurement 51.2 centimeters (made with the top meterstick in Figure A) has three significant figures, and the measurement 51.25 centimeters (made with the lower meterstick) has four significant figures. The right-most digit is always an estimated digit. Only one estimated digit is ever recorded as part of a measurement. It would be incorrect to report the length of the table (Figure A) is 51.253 centimeters as measured with the lower meterstick. This five-significant-figure value would have two estimated digits (the 5 and 3) and would be incorrect because it indicates a *precision* greater than the meterstick can obtain.

Standard rules have been developed for writing and using significant figures, both in measurements and in values calculated from measurements.

Rule 1
In numbers that do not contain zeros, all the digits are significant.

        EXAMPLES:
                4.1327   five significant figures
                5.14   three significant figures
                369   three significant figures

Rules 2
All zeros between significant digits are significant.

        EXAMPLES:
                8.052   four significant figures
                7059   four significant figures
                306   three significant figures

Rule 3
Zeros to the left of the first nonzero digit serve only to fix the position of the decimal point and are not significant.

        EXAMPLES:
                0.0068   two significant figures
                0.0427   three significant figures
                0.0003506   four significant figures

Rule 4
In a number with digits to the right of the decimal point, zeros to the right of the last non-zero digit are significant.

        EXAMPLES:
                53   two significant figures
                53.0   three significant figures
                53.00   four significant figures
                0.00200   three significant figures
                0.70050   five significant figures

Rule 5
In a number that has no decimal point and that ends in one or more zeros (such as 3600), the zeros that end the number may or may not be significant.

The number is ambiguous in terms of significant figures. Before the number of significant figures can be specified, further information is needed about how the number was obtained. If it is a measured number, the zeros are not significant. If the number is a defined or counted number, all the digits are significant.

Confusion is avoided when numbers are expressed in scientific notation. All digits are taken to be significant when expressed this way.

EXAMPLES:

4.6×10^{-5}	two significant figures
4.60×10^{-5}	three significant figures
4.600×10^{-5}	four significant figures
2×10^{-5}	one significant figures
3.0×10^{-5}	two significant figures
4.00×10^{-5}	three significant figures

Rounding

Calculators often display eight or more digits. How do you round such a display to, say, three significant figures? Three rules govern the process of deleting unwanted (insignificant) digits from a calculator number.

Rule 1

If the first digit to the right of the last significant figure is less than 5, that digit and all the digits that follow it are simply dropped.

EXAMPLE:

51.234 rounded to three significant figures becomes 51.2.

Rule 2

If the first digit to be dropped is a digit greater than 5, or if it is a 5 followed by a digit other than zero, the excess digits are dropped and the last retained digit is increased in value by one unit.

EXAMPLE:

51.35, 51.359, and 51.3598 rounded to three significant figures all become 51.4.

Rule 3

If the first digit to be dropped is a 5 not followed by any other digit, or if it is a 5 followed only by zeros, an odd-even rule is applied.

That is, if the last retained digit is even, its value is not changed, and the 5 and any zeros that follow are dropped. But if the last digit is odd, its value is increased by one. The intention of this odd-even rule is to average the effects of rounding off.

EXAMPLES:

74.2500 to three significant figures becomes 72.2
89.3500 to three significant figures becomes 89.4.

Significant Figures and Calculated Quantities

Suppose that you measure the mass of a small wooden block to be 2 grams on a balance, and you find that its volume is 3 cubic centimeters by dipping it beneath the surface of water in a graduated cylinder. The density of the piece of wood is its mass divided by its volume. If you divide 2 by 3 on your calculator, the reading on the display is 0.6666666. However, it would be incorrect to report that the density of the block of wood is 0.6666666 gram per cubic centimeter. To do so would be claiming a higher degree of precision than is warranted. Your answer should be rounded off to a sensible number of significant figures.

The number of significant figures allowable in a calculated result depends on the number of significant figures in the measured data, and on the type of mathematical operation(s) used in calculating. There are separate rules for multiplication and division, and for addition and subtraction.

179

Multiplication and Division

For multiplication and division an answer must have the number of significant figures found in the number with the fewest significant figures. For the density example given above, the answer must be rounded off to one significant figure, 0.7 gram per cubic centimeter. If the mass were measured to be 2.0 grams, and if the volume were still taken to be 3 cubic centimeters, then the answer must still be rounded off to 0.7 gram per cubic centimeter. If the mass were measured to be 2.0 and the volume 3.0 or 3.00 cubic centimeters, the answer must be rounded off to two significant figures: 0.67 gram per cubic centimeter.

Study the following examples. Assume that the numbers being multiplied or divided are measured numbers.

EXAMPLE A:

8.536 x 0.47 = 4.01192 (calculator answer)

The input with the fewest significant figures is 0.47, which has two significant figures. Therefore, the calculator answer 4.01192 must be rounded off to 4.0.

EXAMPLE B:

3840 ÷ 285.3 = 13.45916 (calculator answer)

The input with the fewest significant figures is 3940, which has three significant figures. Therefore, the calculator answer 13.45916 must be rounded off to 13.5.

EXAMPLE C:

36.00 ÷ 3.000 = 12 (calculator answer)

Both inputs contain four significant figures. Therefore, the correct answer must also contain four significant figures, and the calculator answer 12 must be written as 12.00. In this case the calculator gave too few significant figures.

Addition and Subtraction

For addition and subtraction the answer should not have digits beyond the last digit position common to all the numbers being added or subtracted. Study the following examples:

EXAMPLE A:

```
   34.6
   18.8
+  15
   68.4 (calculator answer)
```

The last digit position common to all numbers is that just to the left of where a decimal point is placed or might be placed. Therefore, the calculator answer of 68.4 must be rounded off to 68.

EXAMPLE B:

```
   20.02
   20.002
+  20.0002
   60.0222 (calculator answer)
```

The last digit position common to all numbers is the hundredths place. Therefore, the calculator answer of 60.0222 must be rounded off to 60.02.

EXAMPLE C:

$$345.56$$
$$-\ \underline{245.5}$$
$$100.06 \text{ (calculator answer)}$$

The last digit position common to both numbers in this subtraction is the tenths place. Therefore, the answer must be rounded off to 100.1.

Percentage Error

A measured value is best compared to an accepted value by the percentage of difference rather than the size of the difference. Measuring the length of a 10-centimeter pencil to +/- one centimeter is quite a bit different from measuring the length of a 100-meter track to the same +/- centimeter. The measurement of the pencil shows a relative uncertainty of 10%. The track measurement is uncertain by only 1 part in 10,000, or 0.01%.

The relative uncertainty or relative margin of error in measurements, when expressed as a percentage, is the *percentage of error*. It tells by what percentage a quantity differs from a known accepted value as determined by skilled observers using high-precision equipment. It is a measure of the *accuracy* of the method of measurement, which includes the skill of the person making the measurement. The percentage of error is equal to the difference between the measured value and the accepted value of a quantity divided by the accepted value, and then multiplied by 100. (Note: % error should always be presented as a positive number. Take the absolute value if necessary.)

$$\% \text{ error} = \frac{(\text{accepted value} - \text{measured value})}{(\text{accepted value})} \times 100$$

For example, suppose that the measured value of the acceleration of gravity is found to be 9.44 m/s^2. The accepted value is 9.81 m/s^2. The percentage error is

$$\% \text{ error} = \frac{(9.81 \text{ m/s}^2 - 9.44 \text{ m/s}^2)}{(9.81 \text{ m/s}^2)} \times 100 = 3.77\%$$

Appendix B
Graphing

Many quantities are dependent on one another. The circumference of a circle, for example, depends on the circle's diameter, and vice versa. Tables, equations, and graphs show how dependent quantities are related. Investigating relationships between dependent quantities make up much of the work of physical science. Tables, equations, and graphs are important physical science tools.

Tables

Tables give values of dependent variables in list form. Table A, for example, lists the instantaneous speed of an object relative to its elapsed time of motion. A table organizes experimental data and helps the investigator to deduce relationships.

	Elapsed Time (seconds)	Instantaneous Speed (meters/second)
Table A	0	0
	1	10
	2	20
	3	30
	4	40
	.	.
	.	.
	.	.
	t	$10t$

The relationship between two or more dependent variables can be described using words or can be more concisely expressed using an *equation*. For example, to say that the speed of free fall depends on acceleration and time is said concisely by the equation $v = v_o + gt$, where v_o is the initial speed when time $t = 0$. When free fall begins from $v_o = 0$, we have just $v = gt$.

A *graph* is a pictorial representation of the relationship between dependent variables, such as those found in an equation. By looking at the shape of a graph, we can quickly tell a lot about how the variables are related. For this reason, graphs can help clarify the meaning of an equation or table of numbers. Also, a graph can help reveal the relationship between variables when the equation is not already known. Experimental data is often graphed for this reason.

The most common and simplest graph is the Cartesian graph, where values of one variable are represented on the vertical axis, called the y-axis, and values of the other variable are represented on the horizontal or x-axis. If variable y is *directly proportional* to variable x then the curve will appear to rise from left to right (Figure A). Immediately you can tell that as x increases, y also increases. If, however, x is *inversely proportional* to y then the curve will appear to decrease from left to right (Figure B). In this case, as x increases, y decreases.

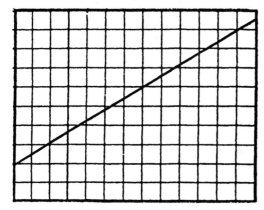

Figure A

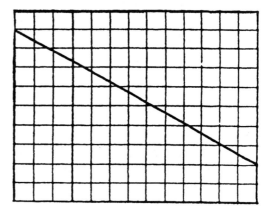

Figure B

Direct and inverse proportionalities are types of linear relationships. Linear relationships have straight-line graphs—the easiest kind to interpret. Figure C shows a graph of the equation $v = gt$. Speed v is plotted along the y-axis, and time t along the x-axis. As you can see, there is a linear relationship between v and t. Here v increases in direct proportion to t; double t and v doubles, triple t and v triples, and so on.

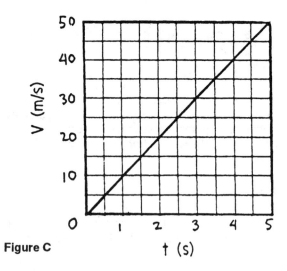

Figure C

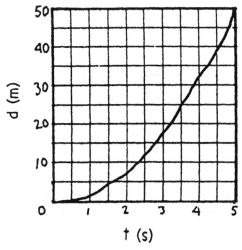

Figure D

Figure D shows a graph of distance d compared to time t in the equation $d = 1/2\, gt^2$. We see by the non-linear curve that d is not directly proportional to t. Interestingly enough, if we make a graph of d versus t^2, a straight line results. This is because distance d *is* directly proportional to t^2, with a slope that is calculated to be 5 m/s^2, or $1/2\, g$ (try it yourself and see). Plotting various powers of non-linear quantities until they form a straight line is one way to find the equation that relates the quantities.

Area Under the Curve

An important feature of a graph that often has physical significance is the area under the curve. Consider the rectangular area bounded by the graph in Figure E. Here the area is the product of the y and x dimensions, speed v and time t respectively (the area of any rectangle is the product of two perpendicular sides). In this case, the area vt has physical significance, for $vt = d$, the distance traveled during the time interval t. Since 50 m/s x 5 s = 250 m, we see the distance traveled in 5 s is 250 m.

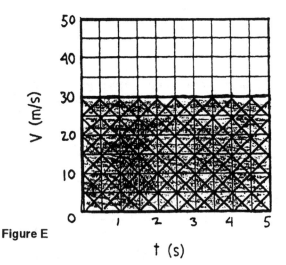

Figure E

The area need not be rectangular. The area beneath any curve of v versus t represents the distance traveled. Similarly, the area beneath a curve of acceleration versus time gives the velocity acquired ($\Delta v = at$), or the area beneath a force versus time curve gives the momentum acquired ($Ft = \Delta mv$). [What does the area beneath a force versus distance curve give?] The area under various curves, including rather complicated ones, can be found by way of an important branch of mathematics—*integral calculus*.

184

Graphing with Conceptual Physical Science

Graphs clarify equations by making them pictorial. You will develop some rudimentary graphing skills in your physical science laboratory. Of particular worth is the lab activity that utilizes a ranging device and a computer, *Sonic Ranger*.

Questions*. Figure F is a graphical representation of the velocity of a ball dropped from a cliff.

1. How long did the ball take to hit the bottom?
2. What was its velocity when it struck bottom?
3. What does the decreasing slope of the graph tell you about the acceleration of the ball with increasing speed?
4. Did the ball reach terminal velocity before hitting the bottom? If so, about how many seconds were required for it to reach its terminal velocity?
5. What is the approximate height of the cliff?
6. How sudden was the impact?

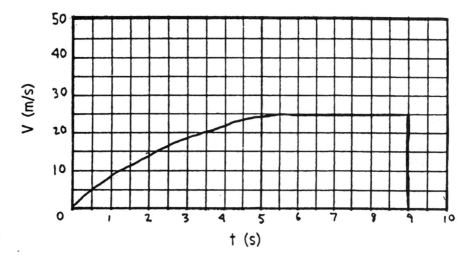

Figure F

Plotting a Graph

The first step in plotting experimental data on a graph is to choose the dimension and scale of the x- and y-axes. Larger graphs allow greater precision. For this reason, the axes should be drawn and labeled so that the data is spaced out over as much of the graph paper as possible. The axes should also be labeled with the units they represent.

Data from a data table is entered into the graph as a series of points. Each point represents the magnitudes—as specified by the axes—of both variable quantities (Figure G). If there is a relation between the variables, then as the data is being plotted, a pattern emerges. In Figure G we see the pattern is a straight line.

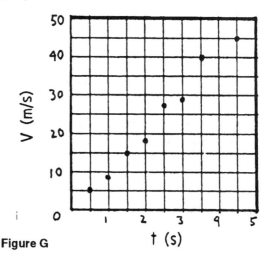

Figure G

---

*Answers: 1) 9 s; 2) 25 m/s; 3) Acceleration decreases as velocity increases (because of air resistance); 4) Yes (because slope curves to zero), about 6 s; 5) Falling distance is nearly 180 m [the area under the curve is about 71 squares—each square represents 2.5 m (5 m/s x 0.5 s = 2.5 m)]; 6) The impact was very sudden as evidenced by the rapid decrease in velocity at the 9th second (what would the graph look like if the ball bounced back up?).

185

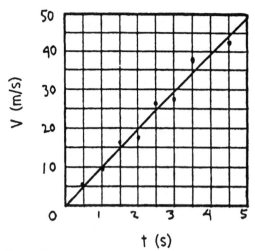

Figure H

Due to experimental error, which includes the limited precision of instruments, data points often deviate from the apparent pattern. For example, data points showing the pattern of a straight line may not fall exactly in a straight-line. In this case a single straight line that approximates all of the data points may be drawn using a ruler or straight edge (Figure H). This straight line need not touch any of the data points for it represents the best fit of all the points together. Non-linear relationships can similarly be sketched with a curve that best fits all the data points.

The *slope* of a linear relationship is found by measuring how high the line on the graph rises compared to how far out it extends. This may be done by drawing a right triangle that has the sloping line as its hypotenuse (Figure I). The slope is equal to the height of the triangle (its "rise") divided by the base (its "run"). Slope = rise/run.

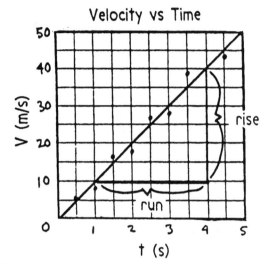

Figure I

Rise is read as the height the right-triangle rises relative to the *y*-axis and run is read as the distance the triangle runs relative to the *x*-axis. The units of the *x*- and *y*-axes must be included. In Figure I, for example, the rise equals 30 m/s and the run equals 3 s. The slope of the line, therefore, equals:

$$\text{slope} = \frac{\text{rise}}{\text{run}} = \frac{30\,\text{m/s}}{3\text{s}} = 10\frac{\text{m/s}}{\text{s}} = 10\frac{\text{m}}{\text{s}^2}$$

The units show us what the slope represents. In this case the sloped line represents the acceleration (m/s^2) of the object. Note the distinction between the *angle* of the line, 45°, and the *slope* of the line, 10 m/s^2.

The final step to creating a graph is assigning a brief but descriptive title, such as "Velocity vs. Time" (Figure I).

186

Appendix C
Conversion Factors

In the physical sciences it is often necessary to convert one unit into another. Units are converted by multiplying the given quantity by a *conversion factor*. A conversion factor is a ratio derived from an equality, hence, a conversion factor always equals 1. Consider the following example:

$$12 \text{ inches} = 1 \text{ foot}$$

A given quantity divided by the same quantity always equals one.

$$\frac{12 \text{ inches}}{12 \text{ inches}} = 1 \qquad \text{or} \qquad \frac{1 \text{ foot}}{1 \text{ foot}} = 1$$

Since 12 inches and 1 foot represent the same quantity, we can perform the following substitutions:

$$\frac{12 \text{ inches}}{1 \text{ foot}} = 1 \qquad \text{or} \qquad \frac{1 \text{ foot}}{12 \text{ inches}} = 1$$

The above two ratios are conversion factors. Since conversion factors are equal to 1, multiplying a quantity by a conversion factor does not change that quantity. While the quantity *doesn't* change, the units *do* change. For example, suppose some item was measured at 60 inches in length. This quantity may be converted into feet by multiplying by the appropriate conversion factor:

$$(60 \text{ inches}) \frac{(1 \text{ foot})}{(12 \text{ inches})} = 5 \text{ feet}$$

$$\Uparrow \qquad\quad \Uparrow \qquad\qquad \Uparrow$$

quantity   conversion   quantity
in inches   factor   in feet

Note how the units of inches cancel each other since they are found in both the numerator and denominator. Similarly, the length of an item measured in feet can be converted to inches using the reciprocal conversion factor.

$$(5 \text{ feet}) \frac{(12 \text{ inches})}{(1 \text{ foot})} = 60 \text{ inches}$$

$$\Uparrow \qquad\quad \Uparrow \qquad\qquad \Uparrow$$

quantity   conversion   quantity
in feet   factor   in inches

To derive a conversion factor, consult a table of unit equalities, such as Table A. The unit you want to obtain in your answer should be set in the numerator of the conversion factor, and the unit you want to cancel should be set in the denominator.

Several conversion factors may be linked together when the relationship between two units is not given directly. For example, suppose you are to convert 5.00 inches into kilometers. In this case you might convert inches into centimeters and then centimeters into meters followed by meters into kilometers.

$$(5.00 \text{ inches}) \frac{(2.54 \text{ cm})}{(1 \text{ inch})} \frac{(1 \text{ meter})}{(100 \text{ cm})} \frac{(1 \text{ km})}{(1000 \text{ meters})} = 0.000127 \text{ kilometers}$$

Alternatively, you might convert inches into feet and then feet into miles followed by miles into kilometers.

$$(5.00 \text{ inches}) \frac{(1 \text{ foot})}{(12 \text{ inch})} \frac{(1 \text{ mile})}{(5280 \text{ feet})} \frac{(1.609 \text{ km})}{(1 \text{ mile})} = 0.000127 \text{ kilometers}$$

The technique of following your units guides you in selecting the proper conversion factor. For example, suppose we wish to convert 15.0 calories per minute into joules per second, which are the units of watts. Then

$$(15.0 \frac{cal}{min}) \frac{(4.187 \text{ J})}{(1 \text{ cal})} \frac{(1 \text{ min})}{(60 \text{ sec})} = 1.22 \frac{J}{sec} = 1.22 \text{ watts}$$

Similarly, we may convert 55.0 miles per hour into meters per second:

$$(55.0 \frac{mi}{hr}) \frac{(5280 \text{ ft})}{(1 \text{ mi})} \frac{(12 \text{ in})}{(1 \text{ ft})} \frac{(2.54 \text{ cm})}{(1 \text{ in})} \frac{(1 \text{ m})}{(100 \text{ cm})} \frac{(1 \text{ hr})}{(3600 \text{ sec})} = 24.6 \frac{m}{sec}$$

In all cases, non-wanted units cancel and only wanted units remain.

Table A Units of Measurement

UNITS OF LENGTH		EXACT
?*		
1 kilometer (km)	= 1000 meters (m)	yes
1 meter (m)	= 1000 millimeters (mm)	yes
	= 100 centimeters (cm)	yes
1 micron (u)	$= 1 \times 10^{-6}$ m	yes
1 nanometer (nm)	$= 1 \times 10^{-9}$ m	yes
1 angstrom (A)	$= 1 \times 10^{-10}$ m	yes
1 inch (in)	= 2.54 cm	yes
1 mile	= 1.6093440 km	yes
	= 5280 feet	yes

UNITS OF MASS AND WEIGHT		
1 kilogram (kg)	= 1000 g	yes
1 gram (g)	= 1000 milligrams (mg)	yes
1 kg	= 2.205 pounds	no
1 pound (lb)	= 453.6 g	no
	= 16 ounces (oz)	yes
	= 4.448 newtons	no

UNITS OF VOLUME		
1 liter (L)	= 1000 milliliters (mL)	yes
1 mL	= 1 cubic centimeter (cc or cm^3)	yes
1 L	= 1.057 quarts	no
1 cubic inch (in^3)	= 16.39 mL	no
1 gallon (gal)	= 4 quarts (qt) = 8 pints (pt)	yes

UNITS OF ENERGY		
1 calorie (cal)	= 4.184 joules (J)	yes
	= amount of heat needed to raise 1 g	
	of water by 1 °C	yes
1 kilocalorie (kcal)	= 1000 cal = 1 nutritional calorie (C)	yes
	= 4.184 kJ	yes
1 joule	$= 1 \times 10^7 \text{ erg} = 1 \text{ kg m}^2/s^2$	yes
1 electron volt (eV)	$= 24.217 \text{ kcal/mol} = 1.602 \times 10^{-19}$ J	no
1 kilowatt hour (kWh)	$= 3.60 \times 10^6$ J	no

UNITS OF PRESSURE		
1 atmosphere (atm)	= 760 mm Hg = 760 torr	yes
	= 14.70 pounds per square inch (psi)	no
	= 29.29 in. Hg	no

*Conversion factors involving exact relationships contain an unlimited number of significant figures (See appendix A).

Exercises:

1. Convert 20.0 calories into joules

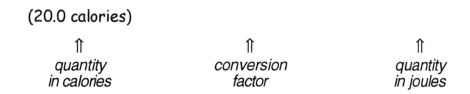

(20.0 calories)

quantity in calories   conversion factor   quantity in joules

2. Convert 26.2 miles into kilometers

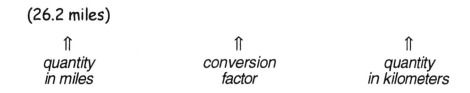

(26.2 miles)

quantity in miles   conversion factor   quantity in kilometers

3. Convert 2.00 miles into centimeters

Answers:

1. Use the conversion factor 4.187 joules/1 calorie to convert 20.0 calories into 83.7 joules.

2. Use the conversion factor 1.609 km/1 mile to convert 26.2 miles into 42.2 kilometers.

3. Convert miles into feet and then feet into inches followed by inches into centimeters to get the answer 322,000 centimeters (See Appendix A for rules concerning significant figures).

Name_____ Period_____ Date_____

FLUIDS: Archimedes' Principle

Sink or Swim

Purpose
In this activity, you will observe the effect of density on various objects placed in water.

Required Equipment and Supplies
equally massive blocks of lead, wood, and Styrofoam
at least two aluminum cans of soda pop—both diet and regular
aquarium tank or sink
egg
wide-mouth graduated cylinder
bowl
salt
spoon
balance

Discussion
Do you know or have you observed in swimming areas that some people have difficulty floating while others don't? This activity should increase your understanding of this disparity.

Procedure
Step 1: Balance equal masses of lead and wood, using a double-beam balance. Repeat, using an equal mass of Styrofoam.

1. How do the *volumes* compare? How do the *densities* therefore compare?

_____

_____

Step 2: Try floating cans of soda pop.

2. Which ones float? Sink?

_____

3. How does the density of the different kinds of soda pop compare to the density of tap water?

_____

_____

4. Hypothesize how the relative densities relate to sugar content.

_____

_____

Step 3: Use a balance to measure the mass of an egg. Using a wide-mouth graduated cylinder, carefully determine the volume of the egg by measuring the volume of water it displaces when it is slowly (gently!) lowered into the graduated cylinder. Calculate its density: $d = m/V$.

Density = _____

Step 4: Now try to float the egg in a bowl of water. Does it float? If not, dissolve enough salt in the water until the egg floats.

5. How does the density of an egg compare to that of tap water?

_____

_____

6. To salt water?

_____

_____

Summing Up

7. Does adding salt to the water make the water less dense or more dense? How?

_____

_____

_____

_____

8. Why do some people find it difficult to float while others don't? What evidence can you cite for the notion that it is easier to float in salt water?

_____

_____

| CONCEPTUAL PHYSICAL SCIENCE **EXPLORATIONS** | **Activity** |

FLUIDS: Archimedes' Principle

Eureka!

Purpose
In this activity, you will investigate the displacement of water by immersed objects.

Required Equipment and Supplies
35-mm film canisters and string
ballast material (nuts, BBs, sand, etc.)
beam balance
500 mL graduated cylinders

Discussion
Archimedes is much remembered for his clever discovery of the method for determining the volume of irregularly shaped objects—like himself! He discovered this while in the public baths of Athens, pondering a problem given to him by the king—how to determine whether a particular gold crown was really 100% gold or not. His discovery is based on a very simple idea that many people don't fully understand. If perchance you are one of them, after doing this activity, you won't be!

Procedure
Your instructor will give you two film canisters, each with contents having different masses, and each with a piece of string attached. Find the mass of each canister.

Mass of lighter canister, m = _____ g

Mass of heavier canister, m = _____ g

Procedure
So that you can easily mark the water level in a graduated cylinder, attach a vertical strip of masking tape along its outer side. Fill the cylinder about three-quarters full of water. Mark the water level on the tape. Submerge the lighter canister and mark the new water level on the tape. Remove the canister.

Predict how the water level for the heavier canister will compare when it is submerged.

_____

Try it and see! Was your prediction correct?

_____

Summing Up
Describe and explain your observations.

_____

_____

CONCEPTUAL PHYSICAL SCIENCE **EXPLORATIONS** | **Experiment**

FLUIDS: Archimedes' Principle

Boat Float

Purpose
In this experiment, you will investigate Archimedes' Principle and the principle of flotation.

Required Equipment and Supplies
spring scale
triple-beam balance
string and masking tape
rock or hook mass
600-mL beaker
500-mL graduated cylinder
clear container or 3-gallon bucket
water
chunk of wood
modeling clay
toy boat capable of carrying a 1200-gram cargo, or 9-inch aluminum cake pan
100-g mass
3 lead masses or lead fishing weights

Discussion
An object submerged in water takes up space and pushes water out of the way. We say the water is *displaced*. Interestingly enough, the water that is pushed out of the way pushes back on the submerged object. For example, if the object pushes a volume of water with a weight of 100 N out if its way, then the water reacts by pushing back on the object with a force of 100 N—Newton's third law. We say that the object is *buoyed* upward with a force of 100 N. This is summed up in Archimedes' principle, which states that the *buoyant force* that acts on any completely or partially submerged object *is equal to the weight of the fluid the object displaces.*

Procedure
Step 1: Use a spring scale to determine the weight of an object (rock or hook mass) that is first out of water and then under water. The difference in weights is the buoyant force. Record.

Weight of object out of water = _____

Weight of object in water = _____

Buoyant force on object = _____

Step 2: Devise a method to find the volume of water displaced by the object. Record the volume of water displaced. Compute the mass and weight of this water. (Remember, 1 mL of water has a mass of 1 g and a weight of 0.01 N.)

Volume of water displaced = _____

Mass of water displaced = _____

Weight of water displaced = _____

1. How does the buoyant force on the submerged object compare with the weight of the water displaced?

_____

_____

NOTE: To simplify calculations for the remainder of this experiment, measure and determine *masses*, without finding their equivalent *weights (W = mg)*. Keep in mind, however, that an object floats because of a buoyant *force*. This force is due to the *weight* of the water displaced.

Step 3: Measure the mass of a piece of wood with a beam balance, and record the mass in Table A. Measure the volume of water displaced when the wood floats. Record the volume and mass of water displaced in Table A.

2. What is the relation between the buoyant force on any floating object and the weight of the object?

_____

_____

3. How does the mass of the wood compare to the mass of the water displaced?

_____

_____

4. How does the buoyant force on the wood compare to the weight of water displaced?

_____

_____

Step 4: Add a 100-g mass to the wood so that the wood displaces more water but still floats. Measure the volume of water displaced and calculate its mass, recording them in Table A.

5. How does the buoyant force on the wood with its 100-g load compare to the weight of water displaced?

_____

_____

Step 5: Roll the clay into a ball and find its mass. Measure the volume of water it displaces after it sinks to the bottom of a graduated cylinder. Calculate the mass of water displaced. Record all volumes and masses in Table A.

6. How does the mass of water displaced by the clay compare to the mass of the clay out of the water?

_____

_____

7. Is the buoyant force on the submerged clay greater than, equal to, or less than its weight out of the water? What is your evidence?

_____

_____

Step 6: Retrieve the clay from the bottom, and mold it into a shape that allows it to float. Sketch or describe this shape. Measure the volume of water displaced by the floating clay. Calculate the mass of the water, and record in Table A.

Table A

Object	Mass (g)	Volume of water displaced (mL)	Mass of water displaced (g)
Wood			
Wood + 50-g mass			
Clay ball			
Floating clay			

Summing Up

8. Does the clay displace more, less, or the same amount of water when it floats as it did when it sank?

_____

_____

9. Is the buoyant force on the floating clay greater than, equal to, or less than the weight of the clay?

_____

_____

10. What can you conclude about the weight of an object and the weight of water displaced by the object when it floats?

_____

_____

11. Is the buoyant force on the clay ball greater when it is submerged near the bottom of the container or when it is submerged near the surface? What is your evidence?

_____

_____

12. Is the pressure that the water exerts on the clay ball greater near the bottom of the container than when submerged near the surface? Exaggerate the depth involved and cite expected evidence.

_____

13.   Why are your last two answers different?

_____

Going Further

Step 7: Suppose you are on a ship in a canal lock. If you throw a ton of lead bricks overboard from the ship into the canal lock, will the water level in the canal lock go up, down, or stay the same? Write down your prediction *before* you proceed to Step 8.

 Prediction for water level in canal lock: _____

Step 8: Float a toy boat loaded with lead "cargo" in a relatively deep container filled with water (deeper than the height of the lead masses). For observable results, the size of the container should be just slightly bigger than the boat. Mark and label the water levels on masking tape placed on the container and on the sides of the boat. Remove the masses from the boat and put them in the water. Mark and label the new water levels.

14.   What happens to the water level *on the side of the boat* when you remove the cargo? What does this say about the amount of cargo carried by ships that float high in the water?

_____

15.   What happens to the water level *in the container* when you place the cargo in the water? Explain why this happens.

_____

_____

16.   Similarly, what happens to the water level in the canal lock when the bricks are thrown overboard?

_____

17.   Suppose the freighter is carrying a cargo of Styrofoam instead of bricks. What happens to the water level in the canal lock if the Styrofoam (which floats in water) were thrown overboard?

_____

18.   When a ship is launched at a shipyard, what happens to the sea level all over the world— no matter how imperceptibly?

_____

19.   When a ship in the harbor launches a little rowboat from the dock, what happens to the sea level all over the world—no matter how imperceptibly?

_____

20.   Cite evidence to support your (different?) answers to 18 and 19.

_____

198

CONCEPTUAL PHYSICAL SCIENCE **EXPLORATIONS** | **Experiment**

FLUIDS: Force and Pressure

Tire Pressure and 18-Wheelers

Purpose
In this experiment, you will investigate the difference between pressure and force.

Required Equipment and Supplies
automobile
owner's manual for vehicle (optional)
graph paper
tire pressure gauge

Discussion
People commonly confuse *force* and *pressure*. Tire manufacturers add to this confusion by saying, "Inflate to 45 pounds," when they really mean, "45 pounds *per* square inch." The pressure on your feet is painfully more when you stand on your toes than when you stand on your whole feet, even though the force of gravity (your weight) is the same. Pressure depends on how a force is distributed. Pressure increases as area decreases, and decreases as area increases.

$$\text{Pressure} = \frac{\text{force}}{\text{area}}$$

Force = pressure x area

Ever wonder why trucks that carry heavy loads have so many tires?
Count the tires on the trucks you see on the highway. Some commonly have 18! Why?

Procedure
Step 1: Position a piece of graph paper directly in front of the front tire of an automobile. Roll the automobile onto the paper.

Step 2: Use a tire gauge to measure the pressure in each tire in pounds per square inch. Record the pressures of each tire. Trace the outline of the tire where it makes contact with the graph paper. Roll the vehicle off the paper. Calculate the area inside the trace in square inches or square cm. Record your data in Table A.

Step 3: Study the tread pattern of the tire. Compare this to the tire mark on the paper. Note that only the rubber in the tread actually presses against the road. The gaps in the tread do not support the vehicle. Estimate what fraction of the tire's contact area is tread and record it in Table A. The area actually in contact with the road is the area inside your trace multiplied by the fraction of tread. Repeat for all four tires.

Step 4: Compute the force each tire exerts against the road by multiplying the pressure of the tire times its area of contact with the road. Record your computations in Table A.

Tire	Area traced on paper (in²)	Estimate of % tread	Area of contact	Pressure (lb/in²)	Force (lb)
Right front					
Left front					
Right rear					
Left rear					

Step 5: Compute the weight of the vehicle by adding the forces from each tire.

Computed weight = _____lb

Is this not quite remarkable—that the weight of a vehicle is equal to the air pressure in its tires multiplied by the area of tire contact? (See if your Uncle Harry knows this!)

Step 6: Ascertain the weight of the vehicle from another source, such as the owner's manual or the local dealer. Sometimes this information is stamped on a plate on the inside of a doorjamb.

Known weight = _____lb

Summing Up

1. How does your computed value of vehicle weight compare to the stated value? What might account for a difference in the actual weight and the weight stated in the owner's manual?

_____

2. When you inflate tires to a higher pressure, what happens to the contact area of the tire against the road surface?

_____

3. A flat tire will read zero on a pressure gauge, when actually air at atmospheric pressure is inside. So a tire gauge is calibrated to measure the air pressure over and above atmospheric pressure. We distinguish between *gauge pressure* and *atmospheric pressure*. Atmospheric pressure is normally 14.7 lb/in² or 10^5 N/m² outside the tire. Gauge pressure is the amount of pressure inside the tire over and above atmospheric pressure. So what is the *total* pressure inside the front tire?

_____

4. Why was the atmospheric pressure (14.7 lb/in²) *not* added to the pressure of the tire gauge when computing the weight of the vehicle?

_____

_____

5. Why do trucks that carry heavy loads have so many wheels—often eighteen?

_____

_____